# Big Data Applications in the Telecommunications Industry

Ye Ouyang
*Verizon Wirless, USA*

Mantian Hu
*Chinese University of Hong Kong, China*

A volume in the Advances in Wireless Technologies and Telecommunication (AWTT) Book Series

www.igi-global.com

Published in the United States of America by
 IGI Global
 Information Science Reference (an imprint of IGI Global)
 701 E. Chocolate Avenue
 Hershey PA 17033
 Tel: 717-533-8845
 Fax: 717-533-8661
 E-mail: cust@igi-global.com
 Web site: http://www.igi-global.com

Library of Congress Cataloging-in-Publication Data

Names: Ouyang, Ye, 1981- editor. | Hu, Mantian, 1981- editor.
Title: Big data applications in the telecommunications industry / Ye Ouyang
 and Mantian Hu, editors.
Description: Hershey, PA : Information Science Reference, [2017]
Identifiers: LCCN 2016044779| ISBN 9781522517504 (hardcover) | ISBN
 9781522517511 (ebook)
Subjects: LCSH: Network performance (Telecommunication)--Reliability. | Big
 data.
Classification: LCC TK5105.5956 .B54 2017 | DDC 384.30285/57--dc23 LC record available at
https://lccn.loc.gov/2016044779

This book is published in the IGI Global book series Advances in Wireless Technologies and Telecommunication (AWTT) (ISSN: 2327-3305; eISSN: 2327-3313)

British Cataloguing in Publication Data
A Cataloguing in Publication record for this book is available from the British Library.

# Advances in Wireless Technologies and Telecommunication (AWTT) Book Series

ISSN:2327-3305
EISSN:2327-3313

## MISSION

The wireless computing industry is constantly evolving, redesigning the ways in which individuals share information. Wireless technology and telecommunication remain one of the most important technologies in business organizations. The utilization of these technologies has enhanced business efficiency by enabling dynamic resources in all aspects of society.

The **Advances in Wireless Technologies and Telecommunication Book Series** aims to provide researchers and academic communities with quality research on the concepts and developments in the wireless technology fields. Developers, engineers, students, research strategists, and IT managers will find this series useful to gain insight into next generation wireless technologies and telecommunication.

## COVERAGE

- Network Management
- Mobile Communications
- Mobile Technology
- Virtual Network Operations
- Radio Communication
- Telecommunications
- Cellular Networks
- Global Telecommunications
- Wireless Broadband
- Grid Communications

IGI Global is currently accepting manuscripts for publication within this series. To submit a proposal for a volume in this series, please contact our Acquisition Editors at Acquisitions@igi-global.com or visit: http://www.igi-global.com/publish/.

# Titles in this Series

*For a list of additional titles in this series, please visit: www.igi-global.com*

*Interference Mitigation and Energy Management in 5G Heterogeneous Cellular Networks*
Chungang Yang (Xidian University, China) and Jiandong Li (Xidian University, China)
Information Science Reference • copyright 2017 • 362pp • H/C (ISBN: 9781522517122)
• US $195.00 (our price)

*Handbook of Research on Advanced Trends in Microwave and Communication Engineering*
Ahmed El Oualkadi (Abdelmalek Essaadi University, Morocco) and Jamal Zbitou (Hassan 1st University, Morocco)
Information Science Reference • copyright 2017 • 716pp • H/C (ISBN: 9781522507734)
• US $315.00 (our price)

*Handbook of Research on Wireless Sensor Network Trends, Technologies, and Applications*
Narendra Kumar Kamila (C. V. Raman College of Engineering, India)
Information Science Reference • copyright 2017 • 589pp • H/C (ISBN: 9781522505013)
• US $310.00 (our price)

*Handbook of Research on Advanced Wireless Sensor Network Applications, Protocols, and Architectures*
Niranjan K. Ray (Silicon Institute of Technology, India) and Ashok Kumar Turuk (National Institute of Technology Rourkela, India)
Information Science Reference • copyright 2017 • 502pp • H/C (ISBN: 9781522504863)
• US $285.00 (our price)

*Self-Organized Mobile Communication Technologies and Techniques for Network Optimization*
Ali Diab (Al-Baath University, Syria)
Information Science Reference • copyright 2016 • 416pp • H/C (ISBN: 9781522502395)
• US $200.00 (our price)

*Advanced Methods for Complex Network Analysis*
Natarajan Meghanathan (Jackson State University, USA)
Information Science Reference • copyright 2016 • 461pp • H/C (ISBN: 9781466699649)
• US $215.00 (our price)

*Emerging Innovations in Wireless Networks and Broadband Technologies*
Naveen Chilamkurti (La Trobe University, Australia)
Information Science Reference • copyright 2016 • 292pp • H/C (ISBN: 9781466699410)
• US $165.00 (our price)

www.igi-global.com

701 E. Chocolate Ave., Hershey, PA 17033
Order online at www.igi-global.com or call 717-533-8845 x100
To place a standing order for titles released in this series,
contact: cust@igi-global.com
Mon-Fri 8:00 am - 5:00 pm (est) or fax 24 hours a day 717-533-8661

# Table of Contents

# Detailed Table of Contents

**Chapter 1**

*Alexis Huet, Nanjing Howso Technology, China*

Development of high-speed LTE connections has induced an increasingly quantity of traffic data over the network. Detection of abnormal traffic from this continuous stream of data is crucial to identify technical problem or fraudulent intrusion. Unsupervised learning methods can automatically describe structure of the data and deduce patterns of the network. There are useful to identify unexpected behaviors and to promptly detect new type of anomalies. In this article, traffic in wireless network is analyzed through different unsupervised models. Emphasis is given on models combining traffic data with time stamps information. A model called Gaussian Probabilistic Latent Semantic Analysis (GPLSA) is introduced and compared with other methods such as time-dependent Gaussian Mixture Models (time-GMM). Efficiency and robustness of those models are compared, using both sampled and LTE traffic data. Those experimental results suggest that GPLSA can provide a robust and early detection of anomalies, in a fully automatic, data-driven solution.

**Chapter 2**

*Yan Wang, Xidian University, China*
*Zhensen Wu, Xidian University, China*

Using the large amount of data collected by mobile operators to evaluate network performance and capacity is a promising approach developed in the recent last years. One of the challenge is to study network accessibility, based on statistical models and analytics. In particular, one aim is to identify when mobile network becomes congested, reducing accessibility performance for users. In this paper, a new analytic methodology to evaluate wireless network accessibility performance through traffic measurements is provided. The procedure is based on ensemble clustering of network cells and on regression models. It leads to identification of zones where the accessibility remains high. Numerical results show efficiency and relevance of the suggested methodology.

**Chapter 3**

Modeling co-occurrence data generated by more than one processes in network is a fundamental problem in anomaly detection. Co-occurrence data are joint occurrences of pairs of elementary observations from two sets: traffic data in one set are associated with the generating entities (Time) in the other set. Clustering algorithms are valuable because they can obtain the insights from the varied distribution associated with generating entities. This chapter leverages co-occurrence data that combine traffic data with time, and compares Gaussian probabilistic latent semantic analysis (GPLSA) model to a Gaussian Mixture Model (GMM) using temporal network data. Experimental results support that GPLSA holds better promise in early detection and low false alarm rate with low complexity of implementation in a fully automatic, data-driven solution.

**Chapter 4**

Feedback data directly collected from users are a great source of information for telecom operators. They are usually retrieved as complaints and survey data. For the mobile telecoms sector, one purpose is to manage those data to identify network problems leading to customer dissatisfaction. In this paper, a quantitative methodology is used to predict dissatisfied users. It focuses on extraction and selection of predictive features, followed by a classification model. Two sets of data are used for experiments: one is related to complaints, the other to survey data. Since the methodology is similar for those two sets, prediction efficiency and influence of features are compared. Specific influence of user loyalty in survey data is highlighted. Thus, the methodology presented in this article provides a reference for the mobile operators to improve procedures for collecting feedback answers.

## Chapter 5
*Mantian (Mandy) Hu, The Chinese University of Hong Kong, China*

Companies have long realized the value of targeting the right customer with the right product. However, this request has never been so inevitable as in the era of big data. Thanks to the tractability of the customers' behavior, the preference information for each individual is collected and updated by the firm in a timely fashion. In this study, we developed a targeting strategy for telecommunication companies to facilitate the adoption of 4G technology. Utilizing the most up to date machine learning technique and the information about individual's local network, we set up a prediction model of consumer adoption behavior. We then applied the model to the real world and conduct field experiment. We worked with the largest telecommunication company in China and used Apache Spark to analyze the data from the complete customer based of a 2nd tie city in eastern China. In the experiment group, we asked the company to use the list we generated as the targets and in the control group, the company used the existing targeting strategy. The results demonstrated the effectiveness of the proposed approach comparing to existing models.

## Chapter 6
*Mantian (Mandy) Hu, The Chinese University of Hong Kong, China*

In the age of Big Data, the social network data collected by telecom operators are growing exponentially. How to exploit these data and mine value from them is an important issue. In this article, an accurate marketing strategy based on social network is proposed. The strategy intends to help telecom operators to improve their marketing efficiency. This method is based on mutual peers' influence in social network, by identifying the influential users (leaders). These users can promote the information diffusion prominently. A precise marketing is realized by taking advantage of the user's influence. Data were collected from China Mobile and analyzed. For the massive datasets, the Apache Spark was chosen for its good scalability, effectiveness and efficiency. The result shows a great increase of the telecom traffic, compared with the result without leader identification.

**Chapter 7**

This chapter focuses on two kinds of targeting in mobile industry: to target churning customers and to target potential customers. These two targeting strategies are very important goals in Customer Relationship Management (CRM). In the first part of the chapter, the author reviews churn prediction models and its applications. In the second part of the chapter, traditional innovation diffusion models are reviewed and agent-based models are explained in detail. Customers in telecom industry are usually connected by large and complex networks. To understand how network effects and consumer behaviors – such as churning and adopting – interplays with each other is of great significance. Therefore, detailed examples are given to network-based targeting analysis.

**Chapter 8**

This chapter is an introduction to multi-cluster based anomaly detection analysis. Various anomalies present different behaviors in wireless networks. Not all anomalies are known to networks. Unsupervised algorithms are desirable to automatically characterize the nature of traffic behavior and detect anomalies from normal behaviors. Essentially all anomaly detection systems first learn a model of the normal patterns in training data set, and then determine the anomaly score of a given testing data point based on the deviations from the learned patterns. The initial step of learning a good model is the most crucial part in anomaly detection. Multi-cluster based analysis are valuable because they can obtain the insights of human behaviors and learn similar patterns in temporal traffic data. The anomaly threshold can be determined by quantitative analysis based on the trained model. A novel quantitative "Donut" algorithm of anomaly detection on the basis of model log-likelihood is proposed in this chapter.

**Chapter 9**

Continuous-Time Markov Chain-Based Reliability Analysis for Future

*Hasan Farooq, University of Oklahoma, USA*
*Md Salik Parwez, University of Oklahoma, USA*
*Ali Imran, University of Oklahoma, USA*

It is anticipated that the future cellular networks will consist of an ultra-dense deployment of complex heterogeneous Base Stations (BSs). Consequently, Self-Organizing Networks (SON) features are considered to be inevitable for efficient and reliable management of such a complex network. Given their unfathomable complexity, cellular networks are inherently prone to partial or complete cell outages due to hardware and/or software failures and parameter misconfiguration caused by human error, multivendor incompatibility or operational drift. Forthcoming cellular networks, vis-a-vis 5G are susceptible to even higher cell outage rates due to their higher parametric complexity and also due to potential conflicts among multiple SON functions. These realities pose a major challenge for reliable operation of future ultra-dense cellular networks in cost effective manner. In this paper, we present a stochastic analytical model to analyze the effects of arrival of faults in a cellular network. We exploit Continuous Time Markov Chain (CTMC) with exponential distribution for failures and recovery times to model the reliability behavior of a BS. We leverage the developed model and subsequent analysis to propose an adaptive fault predictive framework. The proposed fault prediction framework can adapt the CTMC model by dynamically learning from past database of failures, and hence can reduce network recovery time thereby improving its reliability. Numerical results from three case studies, representing different types of network, are evaluated to demonstrate the applicability of the proposed analytical model.

**Chapter 10**

Spectral Efficiency Self-Optimization through Dynamic User Clustering and

*Md Salik Parwez, University of Oklahoma, USA*
*Hasan Farooq, University of Oklahoma, USA*
*Ali Imran, University of Oklahoma, USA*
*Hazem Refai, University of Oklahoma, USA*

This paper presents a novel scheme for spectral efficiency (SE) optimization through clustering of users. By clustering users with respect to their geographical concentration we propose a solution for dynamic steering of antenna beam, i.e., antenna azimuth and tilt optimization with respect to the most focal point in a cell that would maximize overall SE in the system. The proposed framework thus introduces the notion of elastic cells that can be potential component of 5G networks. The proposed scheme decomposes large-scale system-wide optimization problem into small-scale local sub-problems and thus provides a low complexity solution for dynamic system wide optimization. Every sub-problem involves clustering of users to determine focal point of the cell for given user distribution in time and space, and determining new values of azimuth and tilt that would optimize the overall system SE performance. To this end, we propose three user clustering algorithms to transform a given user distribution into the focal points that can be used in optimization; the first is based on received signal to interference ratio (SIR) at the user; the second is based on received signal level (RSL) at the user; the third and final one is based on relative distances of users from the base stations. We also formulate and solve an optimization problem to determine optimal radii of clusters. The performances of proposed algorithms are evaluated through system level simulations. Performance comparison against benchmark where no elastic cell deployed, shows that a gain in spectral efficiency of up to 25% is possible depending upon user distribution in a cell.

# Preface

This book targets at publishing high-quality research, surveys, and case studies in the space of telecommunication networks and markets, leveraging data mining, statistics analysis, data analytics, and machine learning. The articles represented in this book cover a wide variety of topics related to telecommunication and network, big data, machine learning and date mining from a diversity of disciplinary viewpoints, including computer science, electrical engineering, statistics, and management. The studies in this book can be beneficial to researchers and practitioners for solving complex problems and synthesizing knowledge.

In Chapter 1, "Detecting Abnormal Traffic in Wireless Networks Using Unsupervised Models", Alexis Huet discusses detection of abnormal traffic from continuous stream of data over wireless network, and leverages unsupervised learning method to describe structure of data and deduce patterns of the wireless network. The author identifies unexpected behaviors and detects new type of anomalies by combining traffic data with time stamps information and analyzing traffic in wireless network through different unsupervised models. He presents a new analysis model and compares it with other methods. The preliminary results show that the new model can provide a robust and early detection of anomalies as a fully automatic and data-driven solution.

Chapter 2 is titled "Evaluating Wireless Network Accessibility Performance via Clustering-Based Model" by Yan Wang and Zhensen Wu. The authors present a novel analytic methodology to evaluate the performance of wireless network accessibility through traffic measurement, network cells clustering, and regression models. The model leads to identification of where the accessibility remains high and when mobile network becomes congested.

In Chapter 3, "Modeling for Time Generating Network: An Advanced Bayesian Model", Yirui Hu discusses modeling co-occurrence data in anomaly detection. The author presents and compares the Gaussian probabilistic latent semantic analysis (GPLSA) model to a Gaussian Mixture Model (GMM) using temporal network data. Simulation results indicate that this model performs very well compared with other well-known models.

In Chapter 4, "Identifying Dissatisfied 4G Customers from Network Indicators: A Comparison Between Complaint and Survey Data", Xinling Dai analyzes feedback data directly collected from mobile users, to identify network problems which lead to customer dissatisfaction, and further provide references for the mobile operators to improve procedures for collection feedback answers. The author utilizes complaints data and survey data to create a classification model, and presents a quantitative methodology to predict dissatisfied users.

Chapter 5 is written by Mantian (Mandy) Hu with The Chinese University of Hong Kong in Hong Kong. The paper presents a targeting strategy for telecommunication companies to facilitate the adoption of 4G technology. The author applies machine learning technique and leverages information about individual's local network to create a prediction model for consumer adoption behavior. The author implements and verifies the model in real-world scenario.

Chapter 6, "Mining of Leaders in Mobile Telecom Social Networks", is also written by Mantian (Mandy) Hu from The Chinese University of Hong Kong. Her study seeks to exploit social network data collected by telecom operators and mine value from them. The author presents a marketing strategy to help telecom operators to improve marketing efficiency based on social network. This method utilizes influence of mutual peers in social network and identifies the influential users (leaders) who promote the information diffusion prominently.

In "Network-Based Targeting: Big Data Application in Mobile Industry", Chapter 7, Chu Dang investigates targeting in mobile industry to understand how network effects and consumer behaviors- such as churning and adopting- interplays with each other. The author provides not only descriptions of churn prediction models, traditional innovation diffusion model and agent-based models, but also various applications implemented on them.

In Chapter 8, "Anomaly Detection in Wireless Networks: An introduction to Multi-Cluster Technique", Yirui Hu proposes a novel quantitative algorithm of anomaly detection in wireless network based on model log-likelihood. The author applies multi-cluster based analysis to obtain the insights of human behaviors, and learn similar patterns in temporal traffic data, and further leverages the algorithm to automatically characterize the nature of traffic behavior and detect anomalies from normal behaviors.

In Chapter 9, "Continuous-Time Markov Chain-Based Reliability Analysis for Future Cellular Networks", Hasan Farooq, Md Salik Parwez, and Ali Imran design a stochastic analytical model to analyze the effects of faults in a cellular network, and reduce network recovery time for improving the network reliability. The authors model the reliability behavior of a base station, and propose an adaptive fault predictive framework which can adapt the analytical model by dynamically learning from past database of failures. Numerical results are evaluated to demonstrate the applicability of the analytical model.

Md Salik Parwez, Hasan Farooq, Ali Imran, and Hazem Refai present a novel scheme for spectral efficiency optimization through users clustering in their article in Chapter 10, "Spectral Efficiency Self-Optimization Through Dynamic User Clustering and Beam Steering". They decompose large-scale system-wide optimization problem into small-scale local sub-problems and thus provide a low complexity solution for dynamic system wide optimization. Every sub-problem involves clustering of users to determine focal point of the cell for given user distribution in time and space, and determining new values of azimuth and tilt that would optimize the overall system spectral efficiency performance. They propose three algorithms of user clustering to transform a given user distribution into the focal points that can be used in optimization: the algorithm based on received signal to interference ratio (SIR) at the user, the algorithm based on received signal level at the user, and the one based on relative distances of users from the base stations.

We would like to thank Mr. Zhongyuan Li and Ms. Claudia Woo, as well as the many authors and reviewers for their contribution to the articles in this book. The articles in this text span a great deal more of cutting edge areas that are truly interdisciplinary in nature. We hope that you will enjoy reading this book and find the articles informative.

# Chapter 1
# Detecting Abnormal Traffic in Wireless Networks Using Unsupervised Models

**Alexis Huet**
*Nanjing Howso Technology, China*

## ABSTRACT

*Development of high-speed LTE connections has induced an increasingly quantity of traffic data over the network. Detection of abnormal traffic from this continuous stream of data is crucial to identify technical problem or fraudulent intrusion. Unsupervised learning methods can automatically describe structure of the data and deduce patterns of the network. There are useful to identify unexpected behaviors and to promptly detect new type of anomalies. In this article, traffic in wireless network is analyzed through different unsupervised models. Emphasis is given on models combining traffic data with time stamps information. A model called Gaussian Probabilistic Latent Semantic Analysis (GPLSA) is introduced and compared with other methods such as time-dependent Gaussian Mixture Models (time-GMM). Efficiency and robustness of those models are compared, using both sampled and LTE traffic data. Those experimental results suggest that GPLSA can provide a robust and early detection of anomalies, in a fully automatic, data-driven solution.*

DOI: 10.4018/978-1-5225-1750-4.ch001

# INTRODUCTION

Data monitored through telecommunication networks have grown exponentially in the past few years. The resulting flow has become impossible to manually process and analyze. In particular, detection of unexpected traffic behaviors from normal patterns has remain an important issue. This field is critical because anomalies can cause deficiency in network efficiency. Indeed, origin of those anomalies can be a technical problem of a cell or a fraudulent intrusion in the network. There are typically urgent to identify and to fix. Consequently, data-driven systems have been developed to identify anomalies, using machine learning algorithms. The purpose is to automatically extract information from raw data, to identify and alert when an anomaly occurs.

In wireless networks, collected data contain values for different features as well as time stamps. To seek and detect anomalies, values can be modeled and processed using unsupervised algorithms. This kind of algorithms assumes that information about which elements are anomalies is unknown, by using unlabelled data. This is usually the case, since anomalies in the traffic data are rare and may take many forms. Unsupervised algorithms automatically separate and distinguish data structures and patterns. They do not intend to directly detect anomalies, but only to describe and group data. Afterwards, zones of anomalies are deduced from those groups. The main advantage of this methodology is the ability to detect previously unseen or unexpected anomalies.

Another component to take into consideration for wireless networks data is time stamps. This information is commonly collected when data are generated but is not widely used in classic anomaly detection processes. However, network load has daily fluctuations. For example, a normal value during peak period may be an anomaly outside, and remains undetected. Adding time stamp attributes in a model allows to discover periodic behaviors.

In this article, unsupervised models are used to detect anomalies. Specifically, the following sections focus on algorithms combining both values and dates. For this purpose, two models are introduced. The first one is time-GMM, which is a time-dependent extension of GMM (McLachlan & Basford, 1988) by considering each period of time independently. The second one is GPLSA (Hofmann, 1999b), which combines together values and dates processing in a unique machine learning algorithm. This latter algorithm is well-known in text-mining and recommender systems areas, but was barely used in other domains such as anomaly detection. Those algorithms are implemented and tested on sample and traffic data. Their ability to find anomalies and to adapt to new patterns is shown. Robustness, complexity and efficiency of the algorithms are compared.

The rest of the article is organized as follows. In Section II, an overview of techniques to identify anomalies is presented, emphasizing on unsupervised models. In Section III, different unsupervised anomaly detection models are presented. This section insists on two introduced unsupervised models: GPLSA and time-GMM. In Section IV, those models are compared on a sample set to highlight difference of behavior in a simple context. In Section V, computations are performed on real traffic network data and discussed. Finally, in Section VI, conclusion is drawn.

## RESEARCH BACKGROUND

Anomaly detection is a wide topic and a large number of techniques has been used. For a broad overview of those methods, we refer to Chandola, Banerjee, and Kumar (2009).

Research focuses mainly on unsupervised methods to perform anomaly detection (Laskov, Düssel, & Schäfer, 2005; Chawla, Japkowicz, & Kotcz, 2004; Phua, Alahakoon, & Lee, 2004). Most developed are statistical based methods and clustering (Patcha & Park, 2007). Most of the statistical based methods those models are Gaussian model based (Bamnett & Lewis, 1994). Mixture of parametric distributions is also possible, where normal points anomalies correspond to two different distributions (Agarwal, 2007). In clustering methods, the purpose is to separate data points and to group objects which share similarities together. Each group of object is called a cluster. Similarities between objects are usually defined analytically. Many different clustering algorithms exist, differing on how similarities between objects are measured: they can be with some distance measurement, density or statistical distribution. The most popular and simplest clustering technique is K-means clustering (Jain, 2010).

Advanced methods of detection combines statistical hypotheses and clustering, for example with Gaussian Mixture Model (GMM) (McLachlan & Basford, 1988). This method assumes that all the data points are generated from a mixture of K Gaussian distributions. Parameters are usually estimated through an Expectation-Maximization (EM) algorithm, where the aim is to iteratively increase likelihood of the set (Dempster, Laird, & Rubin, 1977). Some studies have used GMM for anomaly detection problems, as described in (Hajji, 2005; Tax & Duin, 1998; Desforges, Jacob, & Cooper, 1998). The number of clusters K to select is not easy. Usually, it is chosen manually and refined after performing different computations for different values. Methods to automatically select a value of K exist, a comparison between different algorithms are presented in (Chiang & Mirkin, 2010).

In telecom traffic data, time stamps are another component to take into consideration. Such information, referred as contextual attributes in Chandola et al. (2009), can dramatically change results of the anomaly detection. Indeed, a value can be considered normal in a certain context (in peak period) but abnormal in another context (in off-peak periods). One way to take into account time stamps is to consider the original GMM model, that is a mixture of K Gaussian distributions, but to weight each distribution differently depending of time. This method was firstly introduced for text-mining (Hofmann, 1999a, 1999b) with a mixture of categorical distributions and named Probabilistic Latent Semantic Analysis (PLSA). Its actual form (with Gaussian distribution) is called GPLSA and is used for recommendation systems (Lu, Pan, & Xiang, 2013). No published article applying GPLSA for anomaly detection has been found.

In the next section, five anomaly detection models for traffic data are presented. The three first models have been already developed and are recalled: Gaussian model, time-dependent Gaussian, GMM. They do not combine together clustering and contextual detection and are expected to have several disadvantages. The two remaining models take into consideration clustering and time stamps: The fourth model is a time-dependent GMM, where a GMM is trained for each time slot independently; The fifth model is Gaussian Probabilistic Latent Semantic Analysis (GPLSA) model, which is solved by optimizing in a unique algorithm all parameters related to clusters and time.

## PRESENTATION OF MODELS

In this section, five different models are defined: Gaussian, time-dependent Gaussian, GMM, time-dependent GMM and GPLSA. The same following notations are used:

- W is traffic data set. This set contains N values indexed with i. N is usually large, that is from one thousand to one hundred million. Each value is a vector of $R^p$, where p is the number of features. Furthermore, each feature is assumed continuous.
- D is time stamps set of classes. This set also contains N values. Since we are expecting a daily cycle, each value $d_i$ corresponds to each hour of the day, consequently stands in $\{0, \ldots, 23\}$.
- X = (W, D) are observed data.
- For clustering methods, we assume that each value is related to a fixed although unknown cluster. The cluster set is named Z. It is a "latent" set since it is initially unknown. We assume that number of clusters K is known.

*Table 1. Example of retrieved traffic data*

| Date | Feat. 1 | ... | Feat. p | W | D |
|---|---|---|---|---|---|
| 04/13 0:00 | 1069 | | 2.4 | (1069, ..., 2.4) | 0 |
| 04/13 0:30 | 1004 | | 2.3 | (1004, ..., 2.3) | 0 |
| ... | ... | | ... | ... | ... |
| 05/04 23:30 | 997 | | 2.7 | (997, ..., 2.7) | 23 |

An example of traffic data retrieved is shown on Table 1.

For each model, the aim is to estimate parameters with maximum-likelihood. When the direct calculation is intractable, EM algorithm is used to find a local optimum (at least) of the likelihood. A usual hypothesis of independence is added, needed to compute likelihood of the product over the set:

- The set of triplets $(W_i, Z_i, D_i)_i$ is an independant vector over the rows i. Note that if the model do not consider D or Z, we can remove this set in this hypothesis.

The different models grouped according to their ability to consider time stamps and clustering are shown on Table 2.

1. **Gaussian Model:** In Gaussian model, the whole data set is assumed to come from a variable following a Gaussian distribution. Consequently, each part of the day has a similar behavior and there is no clusters. Parameters are easily estimated with empirical mean and variance.
2. **Time-Dependent Gaussian Model:** Compared to Gaussian model, a time component is added. Each time of the day is considered independently, following a particular Gaussian distribution. This allows to take into account dependence of time. As for the Gaussian model, parameters are estimated with empirical mean and variance for each class of dates.

*Table 2. Anomaly detection methods compared*

| | No Date | Date |
|---|---|---|
| **No Clustering** | Gaussian | Time-Gaussian |
| **Clustering** | GMM | • Time-GMM<br>• GPLSA |

3. **Gaussian Mixture Model:** Compared to Gaussian model, data is assumed to come from a mixture of Gaussian distributions rather than one Gaussian distribution. The number of cluster K is fixed in advance.

   a. Each record belongs to a cluster $Z_i = k \in \{1,\ldots,K\}$ with probability $\alpha_k$.

   b. Each variable $(W_i \mid Z_i = k)$ follows a Gaussian distribution with mean and variance $m_k$, $\Sigma_k$.

Therefore, each record belongs to an unknown cluster. The task is to estimate both probability to be is each cluster and parameters of each Gaussian distribution. To solve this problem, the following decomposition is done:

$$P(W_i) = \sum_k P(W_i \mid Z_i = k)P\left(Z_i = k\right).$$

The parameters can be successively updated with an EM algorithm.

4. **Time-Dependent Gaussian Mixture Model:** Combining model described in paragraph B and C, we obtain the time-dependent GMM model. This model includes both clustering and time-dependence. As in paragraph C, the EM algorithm is used to estimate parameters.

   a. For each $s \in \{0, \ldots, 23\}$, each record such that $D_i = s$ belongs to a cluster $Z_i = k \in \{1,\ldots,K\}$ with probability $\alpha_{k,s}$.

   b. For each $s \in \{0, \ldots, 23\}$, each variable $(W_i \mid Z_i = k)$ such that $D_i = s$ follows a Gaussian distribution with mean and variance $m_k^s$, $\Sigma_k^s$.

5. **Gaussian Probabilistic Latent Semantic Analysis model:** GPLSA is based on the classic GMM, but introduces a novel link between data values and time stamps. In time-GMM, the different classes of dates are considered and trained independently. On the contrary, GPLSA introduces dependence between latent clusters and time stamps, but only within those two variables. That is, knowing latent cluster Z, there is no more dependence in time. To be computationally tractable, an hypothesis of independence is added between traffic data and time stamps are conditionally independent knowing Z. Explicitly:

   a. For each $s \in \{0, \ldots, 23\}$, each record such that $D_i = s$ belongs to a cluster $Z_i = k \in \{1,\ldots,K\}$ with probability $\alpha_{k,s}$.

   b. Each variable $(W_i \mid Z_i = k)$ follows a Gaussian distribution with mean and variance $m_k$, $\Sigma_k$.

   c. For all i, $P\left(W_i \mid D_i, Z_i\right) = P\left(W_i \mid Z_i\right)$.

To solve this problem, the following decomposition is done (the independent assumption knowing Z is used for the first factor of the sum):

$$P(W_i \mid D_i = s) = \sum_k P(W_i \mid Z_i = k)P(Z_i = k \mid D_i = s).$$

The EM algorithm is straightforward to adapt in this case to iteratively increase likelihood and estimate parameters. Finally, exact update formulas are obtained.

## COMPARISON OF MODELS

All five models defined earlier are applied on a sample set. The purpose is to compare the ability to detect anomalies and to check robustness of the methods. The sample set is built to highlight difference of behavior between models in a simple and understandable context. Consequently, only one feature is considered in addition with time stamp dates.

On this set, we observe that time-GMM and GPLSA are able to detect the anomalies of the set, and are then potential candidates for anomaly detection is a time-dependent context. Furthermore, we show that GPLSA is more robust and allows a higher interpretation level of resulting clusters.

## 1. Sample Definition

The sample is built by superposing the three following random sets:

$$t \mapsto \cos\left(2\pi t \, / \, T\right) + \varepsilon$$

$$t \mapsto \cos\left(\pi + 2\pi t \, / \, T\right) + \varepsilon$$

$$t \mapsto -2.5 + \varepsilon$$

where $\varepsilon$ are independent random variables for each t sampled according to the continuous uniform distribution on [0, 1], and where T is selected to have a daily period. The range of the two first functions is 24 hours, whereas the third one is only defined from 0:00 to 15:00. Three anomalies are added on this set, defined respectively at 6:00, 12:00 and 18:00 with values $-1.25, 0.5$ and $1.65$. The resulting set is shown on Figure 1.

## 2. Anomaly Identification

All five models are trained and likelihood of each point is computed for each model. Since 3 anomalies are expected to be found in this sample set, the 3 lowest likelihood are defined as anomalies for each model. For clustering process, the chosen number of clusters is K = 5.

The results are shown on Figure 1. On (a), the whole data set is modeled as one Gaussian distribution and no expected anomalies are found. On (b), each period is trained with a Gaussian distribution, and only the anomaly at 18:00 is discovered. On (c), the whole set is clustered and only the anomaly at 6:00 is discovered. Finally, on (d) and (e), the time-GMM and GPLSA models are trained and the same results are obtained: the 3 anomalies are successively detected.

Thus, time-GMM and GPLSA are both able to detect expected anomalies on this example, contrary to other methods.

*Figure 1. Anomaly detection for 5 different models on the sample set defined in Section IV. The three values with the lowest likelihood are circled in orange. Each color represents a different time stamp class (only 1 class for (a) and (c); 24 classes for (b) and (d). (a) Gaussian; (b) Time-Gaussian; (c) GMM; (d) Time-GMM and GPLSA*

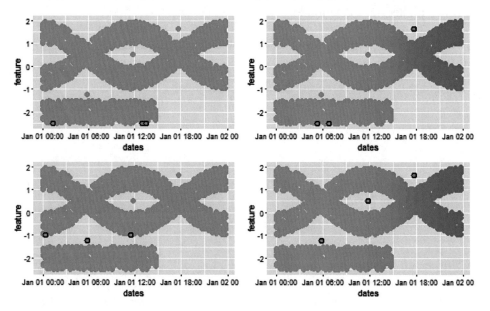

## 3. Comparison between Time-GMM and GPLSA

The same anomalies have been detected with time-GMM and GPLSA. However, the way to detect those anomalies is different, and a comparison is summarized on Table 3.

First, GPLSA considers time stamps and values at once, that is all parameters are estimated at the same time. Consequently, consecutive dates can share similar clustering behaviors. With time-GMM, parameters are trained independently for each class of dates, and no relation exists between clusters of different classes.

Secondly, the number of clusters in each class is soft for GPLSA, i.e. it can be different to the specified number of clusters for some class of dates. This allows to automatically adapt the number of cluster depending of the cluster needed in the model. In time-GMM, each class has a specified number of clusters. This fact can be shown on Figure 2, where the first seven hours is plotted for identified clusters for time-GMM (a) and GPLSA (b).

Thirdly, model is trained with the whole data for GPLSA, whereas only a fraction of data is used for each time-GMM computation. If there is a limited number of data in a class of dates, this behavior can cause failure to correctly estimate time-GMM parameters.

Fourthly, the number of parameters to estimate is $D \times K + 2$ for GPLSA and $2 \times D \times K$ for time-GMM (with D number of classes and K number of clusters). Consequently, there are less parameters to estimate with GPLSA.

On the whole, GPLSA implies a better interpretation level of resulting clusters over time-GMM, combined with a higher robustness.

*Table 3. Comparison between time-GMM and GPLSA*

|  | Time-GMM | GPLSA |
|---|---|---|
| Cluster number | Fixed number of clusters at each date | Number of clusters can adapt to each date |
| Cluster relations | No relation between clusters of each date | Homogeneity of clusters across dates |
| Interpretability | low | high |
| Data used | Only a part of data is used at each date | All data is used for each date |
| Nb. of param. | 2DK | DK+ 2 |
| Robustness | medium | high |

## RESULTS AND DISCUSSION

In this section, anomaly detection is performed on real traffic network data. According to comparison of models done in before, only GPLSA is selected to deduce anomalies. In paragraph A, the collected data set is described and pre-processed. Then, GPLSA is applied and results are shown in paragraph B. This paragraph specifically focuses on observed behavior after applying the algorithm. Finally, paragraph C highlights ability of GPLSA to perform anomaly detection.

## 1. Data Description and Preprocessing

Data have been gathered from a Chinese mobile operator. They comprise a selection of 24 traffic features collected for 3000 cells in the city of Wuxi, China. Those features are only related to cell sites and do not give information about specific users. They represent for example average number of users within cell or total data traffic for the last quarter of hour. The algorithm is trained over two weeks, with one value each quarter of hour for each cell.

Rows of data containing missing values are discarded. Only values and time stamps are taken into consideration for computations, discarding cells identification number. Some features only take nonnegative values and have a skewed behavior. Consequently, some features are pre-processed by applying logarithm. To keep interpretability, we do not apply feature normalization on variables. Finally, we expect that GPLSA can manage this set, even though some properties of the model are not verified, such as normality assumptions.

## 2. Computations and Results

GPLSA model is trained for the feature corresponding to the "average number of users within cell". $K = 3$ clusters are selected. Anomalies are values with the lowest resulting likelihoods, computed to get on average 2 alerts and 8 warnings each day. Visual results are shown on Figure 2.

On (a), the three clusters are identified whereas on (b), a different color is used for each class of dates. On (c), the different log-likelihood values are shown. Finally, on (d), estimation of the probability $\alpha_{k,s}$ to be in each cluster k knowing $D = s$ is plotted.

Anomalies can be shown on (a), (b) and (c). Extreme values related to each class of dates are correctly detected. On (a) and (d), identified clusters are shown in three distinct colors. The probability to be in each cluster varies across class as expected, with a lower probability to be in the upper cluster during off-peak hours. Also, as shown on (a), the upper cluster has a symmetric shape and the mean value is relatively similar across dates.

## DISCUSSION

According to the results, GPLSA is able to detect anomalies in a time-dependent context. Global outliers are identified (for example on Figure 2 (b) at Apr. 15 16:00 in red) as well as context-dependent anomalies (for example at Apr. 15 5:00 in orange). Off-peak periods are taken into consideration, and unusual values specific to those periods are detected.

Gaussian hypothesis on GPLSA is not really constraining. As shown on Figure 2 (a), clusters are adaptable and try to fit Gaussian distributions. There are appropriate to represent values distribution for each class of dates and cluster.

Cluster adaptation is shown on Figure 2 (d). The three clusters are able to represent different level of values. The upper cluster represents higher values, which are more probable during peak periods. The lower cluster represents lower values, with a roughly constant probability. The third cluster in the middle is also useful to obtain a good anomaly detection behavior. Indeed, results are not adequate with $K = 2$ clusters.

About anomaly detection itself, a threshold indicating the number of alerts to deduce can be set. This method of detection is static and relatively simple. Improving this method of detection is possible and straightforward through likelihood computations: inside a cell, an anomaly could be detected with a repetition of low likelihood scores.

*Figure 2. Identified clusters for 2 models on the sample set defined in Section IV between 0:00 and 7:00. On (a), each class of one hour contains 5 clusters, and clusters are not related across hours. On (b), the whole set contains 5 clusters. (a) Time-GMM; (b) GPLSA*

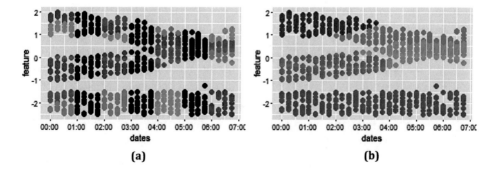

(a)                                            (b)

*Figure 3. Anomaly detection with GPLSA from traffic data set presented in Section V. Plots are restricted to two days on (a), (b) and (c). Red and orange points are related to the lowest likelihoods obtained, with an average of 2 red points and 8 orange points each day. (a) Values as a function of dates, with clusters identified; (b) Values as a function of dates, with classes identified; (c) Log-likelihood of values of the set as a function of dates; (d) Probability to be in a cluster knowing date class*

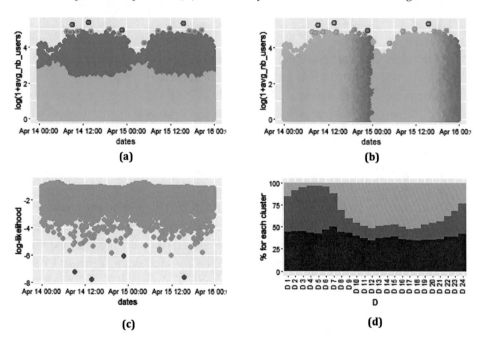

## CONCLUSION

Unsupervised models to detect anomalies in wireless network traffic are presented and compared. GPLSA model is introduced and its robustness, interpretability and ability to detect anomalies are shown, compared to other methods such as time-GMM. Anomaly detection is performed and analyzed in a real traffic data. Adaptability of the GPLSA is highlighted in this context, even those with new patterns and difficult to manually predict. As a result, mobile operators get a versatile way to identify and detect anomalies, reducing the cost of possible aftermaths.

Improvement of the methodology could be operated. From now, once the model is computed, anomaly detection is only based on punctual detection through likelihood values. A dynamic detection from consecutive values of likelihood could be interesting to increase credibility of each alert and reduce number of false alarms. Also, model is only trained from a fixed data set. This could be extended by considering real-time stream data dealt in an on-line fashion. Thus, new patterns should be updated quickly, improving responsiveness and anomaly identification.

# REFERENCES

Agarwal, D. (2007). Detecting anomalies in cross-classified streams: A bayesian approach. *Knowledge and Information Systems*, *11*(1), 29–44. doi:10.1007/s10115-006-0036-4

Bamnett, V., & Lewis, T. (1994). *Outliers in statistical data*. Chichester, UK: John Wiley & Sons.

Chandola, V., Banerjee, A., & Kumar, V. (2009). Anomaly detection: A survey. *ACM Computing Surveys*, *41*(3), 15. doi:10.1145/1541880.1541882

Chawla, N. V., Japkowicz, N., & Kotcz, A. (2004). Editorial: Special issue on learning from imbalanced data sets. *ACM Sigkdd Explorations Newsletter*, *6*(1), 1–6. doi:10.1145/1007730.1007733

Chiang, M. M. T., & Mirkin, B. (2010). Intelligent choice of the number of clusters in k-means clustering: an experimental study with different cluster spreads. *Journal of Classification, 27*(1), 3-40.

Dempster, A. P., Laird, N. M., & Rubin, D. B. (1977). Maximum likelihood from incomplete data via the EM algorithm. *Journal of the Royal Statistical Society. Series B. Methodological, 39*, 1–38.

Desforges, M. J., Jacob, P. J., & Cooper, J. E. (1998). Applications of probability density estimation to the detection of abnormal conditions in engineering. *Proceedings of the Institution of Mechanical Engineers. Part C, Journal of Mechanical Engineering Science, 212*(8), 687–703. doi:10.1243/0954406981521448

Hajji, H. (2005). Statistical analysis of network traffic for adaptive faults detection. *IEEE Transactions on Neural Networks*, *16*(5), 1053–1063. doi:10.1109/TNN.2005.853414 PMID:16252816

Hofmann, T. (1999a). Probabilistic latent semantic indexing. In *Proceedings of the 22nd annual international ACM SIGIR conference on Research and development in information retrieval (SIGIR '99)* (pp. 50-54). New York, NY: ACM. doi:10.1145/312624.312649

Hofmann, T. (1999b). Probabilistic latent semantic analysis. In *Proceedings of the Fifteenth conference on Uncertainty in artificial intelligence* (pp. 289-296). Burlington, MA: Morgan Kaufmann Publishers Inc.

Jain, A. K. (2010). Data clustering: 50 years beyond K-means. *Pattern Recognition Letters*, *31*(8), 651–666. doi:10.1016/j.patrec.2009.09.011

Laskov, P., Düssel, P., Schäfer, C., & Rieck, K. (2005). *Learning intrusion detection: supervised or unsupervised? In Image Analysis and Processing–ICIAP 2005* (pp. 50–57). Berlin: Springer. doi:10.1007/11553595_6

Lu, Z., Zhong, E., Zhao, L., Xiang, E., Pan, W., & Yang, Q. (2013). *Selective Transfer Learning for Cross Domain Recommendation*. Philadelphia, PA: SDM. doi:10.14711/thesis-b1240240

McLachlan, G. J., & Basford, K. E. (1988). *Mixture models: Inference and applications to clustering*. New York: Dekker.

Patcha, A., & Park, J. M. (2007). An overview of anomaly detection techniques: Existing solutions and latest technological trends. *Computer Networks*, *51*(12), 3448–3470. doi:10.1016/j.comnet.2007.02.001

Phua, C., Alahakoon, D., & Lee, V. (2004). Minority report in fraud detection: classification of skewed data. *ACM SIGKDD Explorations Newsletter, 6*(1), 50-59.

Tax, D. M. J., & Duin, R. P. W. (1998). Outlier detection using classifier instability. In *Advances in Pattern Recognition* (pp. 593–601). Berlin: Springer. doi:10.1007/BFb0033283

# Chapter 2

# Evaluating Wireless Network Accessibility Performance via Clustering-Based Model:
## An Analytic Methodology

**Yan Wang**
*Xidian University, China*

**Zhensen Wu**
*Xidian University, China*

## ABSTRACT

*Using the large amount of data collected by mobile operators to evaluate network performance and capacity is a promising approach developed in the recent last years. One of the challenge is to study network accessibility, based on statistical models and analytics. In particular, one aim is to identify when mobile network becomes congested, reducing accessibility performance for users. In this paper, a new analytic methodology to evaluate wireless network accessibility performance through traffic measurements is provided. The procedure is based on ensemble clustering of network cells and on regression models. It leads to identification of zones where the accessibility remains high. Numerical results show efficiency and relevance of the suggested methodology.*

DOI: 10.4018/978-1-5225-1750-4.ch002

## INTRODUCTION

The last few years have witnessed a quick development of wireless connections, together with a growth of smart devices consumers. Those induced a huge increase in the network flow of data. However, the network capacity is limited by infrastructure deployment. An heavy consumption leads to a deterioration of the network accessibility for users. Mobile operators have then to size correctly the network capacity. This capacity cannot be over-sized dues to infrastructure high price, but must be sufficiently large to avoid overloading.

Data collected by mobile operators can be managed and understood to evaluate wireless network accessibility performance. In the traditional method, accessibility is only related to the number of subscribers. Nowadays, this method fails to render the high diversity of traffic patterns and user behaviors. Indeed, consumption of the services becomes more versatile: In addition with classic phone use, there are data-based services such as web browsing, video communication or streaming. Therefore, a critical needs is to dedicate novel methodologies to evaluate accessibility performance. This is done by understanding when the network begins to degrade and to be less accessible.

Relationships between accessibility, capacity management and network performance have been studied, mainly through simulations, both for 3G UMTS (Universal Mobile Telecommunications System) and LTE networks.

About network design, heuristic algorithms (Tsao & Lin, 2002; Szlovencsak, Godor, Harmatos, & Cinkler, 2002) are developed to provide network topologies which ensure a low traffic loss. Also, an analysis of capacity through uplink and downlink is performed in Navaie and Sharafat (2003), resulting in new approaches for network sizing. In Amzallag, Bar-Yehuda, Raz, and Scalosub (2013), an optimization for choosing cells maximizing their use in LTE networks leads to a better usage of network's capacity.

About accessibility, quality of experience prediction models is presented in Khan, Sun and Ifeachor (2012), based on encoded videos. They introduce an adaptation scheme when the accessibility begins to degrade. In Engels et al., (2013), autonomous adjustment of some optimization parameters through time are performed for LTE networks, depending of the network resources. In Ouyang and Fallah (2010) and Ouyang et al., (2014), traffic behavior are simulated through different scenarios to understand the throughput behavior, for both UMTS and LTE networks. In Ouyang, Yan, and Wang (2015), and Ouyang and Yan (2015), this methodology is used again through crowdsourcing-based analytics to evaluate voice or app accessibility. Finally, in Hu, Ouyang, Yao, Fallah and Lu (2014), a relational algorithm between LTE network resources and an wireless network KPI (Key Performance Indicator) is introduced to forecast network resource consumptions.

In this paper, we introduce a new wireless analytic methodology to evaluate accessibility performance. This methodology links network accessibility with traffic measurements performed by the mobile carrier. As in Hu et al., (2014), the network accessibility is depicted with a wireless network KPI, and traffic measurements by network resources. A challenge is then to detect when network starts to deteriorate and to be less accessible.

The first innovation of our procedure is to take into account the non-homogeneous behavior of cells of the network. Cells are not considered individually nor agglutinate, but clustered into groups. This allows to bring out cells which behave similarly. For example, some cells may be constrained to an higher pressure than others, or to a different consumption of the services.

The second innovation is to get an automatic procedure, where each step can be monitored. After selecting features of interest, the whole process is driven by data through machine learning algorithms. It is intended to provide robust and adaptive predictions. For this purpose, an ensemble clustering method is used and an appropriate regression algorithm to predict the KPI is selected.

Then, from selected clusters and obtained prediction functions, we infer when the network begins to degrade. A comfortable zone is deduced as a function of the network resources. This zone indicates when the accessibility performance remains sufficiently high.

This methodology is therefore tested, using data provided by China Mobile.

The paper is organized as follows. The section titled: Methodology to Evaluate Accessibility defines KPI and network resources retrieved and necessary to conduct our methodology. Then, the whole procedure is introduced in the section titled: Detailed Methodology Description. Specifically, we deduce from data a comfortable zone when accessibility remains high. In Section IV, major steps of the methodology are detailed. The results section focuses on numerical results, validation of each step of the procedure and discussion of those results, from the China Mobile data set. Finally, the last section draws conclusion.

# DATA

Data are collected for different network cells on a fixed area. For each cell, recordings of various traffic measurement features are obtained. First, network features data are retrieved. They describe for each cell different variables of the network, such as the congestion ratio or mean time to establish connection. Those features are directly accessible by the network and will be used as predictor variables. Then, a sample data of a KPI related to wireless network accessibility performance is available. This KPI has to be deduced and is our target objective to evaluate accessibility. Therefore,

*Table 1. Data retrieved to conduct our methodology*

| date | FailEstab | KPI | Net. R. 1[a] | ... | Net. R. 40[a] |
|------|-----------|-----|--------------|-----|---------------|
| 7/29 0:00 | 0 | 99.9 | 1022 | | 3.5 |
| 7/29 0:30 | 2 | 96.1 | 1025 | | 3.4 |
| ... | ... | ... | ... | | ... |
| 11/26 10:00 | 0 | 99.8 | 1019 | | 3.8 |

[a.]Network resource

we want to derive predictive functions of this KPI from network resources. For example, this KPI can be the RAB (Radio Access Bearer) Setup Success Ratio of all PS (Packet Switch) services. A third element is necessary to conduct the procedure: The knowledge of number of times the cell causes PS domain failed to establish the number of RAB (called later FailEstab).

Numerical results detailed in Section IV are based on data collected by China Mobile Communications Corporation. Those data have been obtained for analysis by co-operation with this company. They are retrieved on cells located at Xuanwu, a district of Nanjing (China), from July to November 2014. This area is divided in 1668 separated cells. For each cell, every 30 minutes, there is a record of network resources features, the KPI of interest and FailEstab. As a result, the number of recordings for each cell is usually about 5610, but many cells contain missing values. Here, the number of network resources features is 40. As an example, we show on Table 1 a sketch of retrieved data for a particular cell.

## METHODOLOGY TO EVALUATE ACCESSIBILITY

A general description of model and methodology is done in this section. Given each network resource feature, the aim is to identify a zone where the considered KPI remains sufficiently high. Paragraph A describes an initial model which links network resource feature and KPI. This model is set as a reference, but one may keep in mind that data can differ from this initial model and fits have to adjust accordingly. Paragraph B deals with separation of the data set to obtain homogenous components. Two elements modifying shape of the relationship are identified. First, change of the discrete FailEstab value leads to distinct curves. Then, different cells can be related to different behaviors and separation of cells have to be considered. Paragraph C formulates the global methodology to deal with data and evaluate accessibility.

## 1. Model of the Relationship

Behavior of the KPI as a function of the network features follows a specific curve. Basically, we assume that the value of the KPI follows a power law when a network resource x increase. This relationship is shown on (1), for scale parameter a, power parameter b, and constant parameter c. A graphical example of this shape is shown on Figure 2 (b).

$$KPI(x) = a / x^b + c \qquad (1)$$

An interest of (1) is to deduce a linear link taking the logarithm. For example, a linear relation between $\log(x)$ and $\log(c - KPI(x))$ can be exhibited (ensuring that those quantities exist). As mentioned before, the true relationship can differ from (1), therefore some non-linear regressions are used in our methodology to predict the KPI, instead of using linear regressions only.

## 2. Separation of the Data Set

Collected raw data are related to a combination of different behaviors. Then, it is initially difficult to distinguish curves described in (1). To get a functional relationship, separation of the data set into homogenous components is needed.

The first element to take into consideration is the FailEstab network feature. It takes discrete nonnegative values, each of them being related to a specific shape. In the following, data restricted to each value is used to perform computations. On Figure 1 (a), the whole data set is plotted to show a KPI as a function of a network resource. The output looks like a mixture of different power law curves. On Figure 1 (b), this set is restricted to FailEstab = 2, exhibiting a curve closer to equation (1).

The second element is cell id. Each cell has its own behavior and can be related to a particular curve shape. Consequently, to agglutinate all cells together will make difficult the prediction of the KPI. However, to perform computations for each cell is also problematic, since cell may have been accessible for the whole measurement duration. An example of those behaviors is depicted on Figure 2, for data restricted to FailEstab = 2. On Figure 2 (a), all cells are plotted together. On Figure 2 (c), only one specific cell is plotted, but this cell only takes high accessibility values. Therefore, we need to collect a sufficient amount of data for each behavior. Introduced solution will be to cluster cells together into homogeneous groups, as shown on Figure 2 (b).

## 3. Whole Process Methodology

This paragraph intends to describe the global methodology. Further description of each step is available in the Section IV.

The first step is data retrieving and preparation. In this step, data is collected for each cell. Data set is separated into a training (60%), a cross-validation (20%) and a test set (20%). Then, each set is separated according to each value of FailEstab. Also, selection of features allowing to predict the KPI value is done in this step (manually, or by exhaustive search with linear regression (Miller, 2002)).

The second step is to group cells together into homogeneous groups. A clustering process is used for this purpose. The number of groups is not defined in advance. Consequently, at this step, the clustering process is performed for a number of clusters K varying from 1 to 20. Note that for each value of K, the defined groups of cells are identical for all values of FailEstab.

The third step has three parts. First, for each value of K, for each value of FailEstab, for each cluster from 1 to K, a regression process is applied to fit data (on the training set). Then, a summarizing value is defined for each K, taking into account prediction accuracies and difference of behavior of clusters (on the cross-validation set). Those summarizing values allow to select a best number of cluster K. Afterwards, prediction functions are retrieved for this K, for each value of FailEstab and each cluster from 1 to K.

*Figure 1. A KPI as a function of a network resource. On (a), the whole data set is plotted. On (b), the set is restricted to FailEstab = 2.*

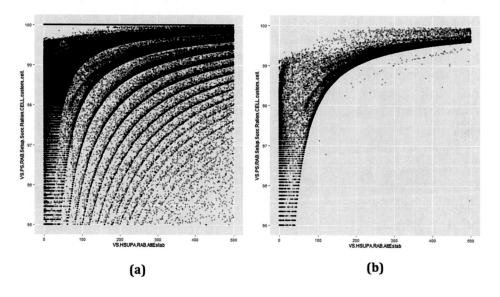

(a)                                        (b)

The fourth step is to deduce a comfortable zone from those prediction functions for each cell and each value of FailEstab. The comfortable zone is defined such that the accessibility remains high.

## DETAILED METHODOLOGY DESCRIPTION

In section III, the whole methodology has been introduced. In the current section, major steps of the methodology are described. First, clustering process is carried out with an ensemble clustering algorithm. Then, regression process combines different regression computations to fit data. Finally, deduction of a comfortable zone is explained.

## 1. Clustering Process of Cells

Through a clustering process, similar cells are let into groups. For now, computations are performed for a number of clusters K varying from 1 to 20. Three steps are necessary: First, for each cell, extraction of summarizing values are deduced from the data set; Then, each cell being summarized, cells are set together using different statistical clustering algorithms; Finally, clustering results are put together into a final clustering, involving an ensemble clustering algorithm.

In the first step, values are extracted from training data set to summarize each cell. We expect a relation as shown on (1) when plotting the KPI values as a function of each selected feature. Therefore, taking the logarithm, a slightly linear relationship exists. Performing a linear regression, we extract intercept and slope for each selected feature, for the first values of FailEstab (limiting to 99% of the data

*Figure 2. A KPI as a function of a network resources for data restricted to FailEstab = 2, plotted for the indicated cells. (a) all cells; (b) clustered cells; (c) one cell*

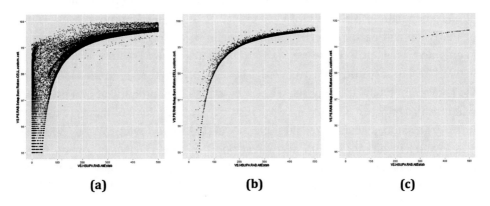

(a)　　　　　　　　(b)　　　　　　　　(c)

points). Those intercept and slope values summarizes the cell, but may contain redundant information. Then, a principal component analysis is done to retrieve 99% of the information from those extracted values. Resulting values contain relevant information of the cell behavior.

In the second step, multiple clustering algorithms are performed. The choice to compute more than one clustering arises from the fact that each particular clustering algorithms have some limitations and are usually specialized to a single task. Consequently, to overcome those limitations, 6 clustering algorithms are carried out. They are chosen to cover a broad spectrum of clustering methods. Those are: k-means (MacQueen, 1967), k-medoids (Kaufman & Rousseeuw, 1987), fuzzy C-means (Bexdek, 2013), CLARA clustering (Kaufman & Rousseeuw, 2009), Gaussian mixture model (GMM) (McLachlan & Basford, 1988) and agglomerative hierarchical clustering.

In the third step, an ensemble clustering is deduced from those six clustering methods. This ensemble clustering tries to find a single consensus from different clusterings. The algorithm used is the co-association matrix based method defined in (Fred, 2001), detailed as follows. First, for each clustering method, computation of the co-association matrix is performed. This matrix is a square matrix and the number of rows is the number of cells. It is defined such that (i, j) is let with 1 if i and j belong to the same cluster, and 0 otherwise. Then, from those matrices, a consensus matrix is deduced. This consensus matrix is defined as the mean of all seven matrices. The final consensus partition is generated by taking a threshold of 0.5 on the consensus matrix. In other words, i and j belong to the same final consensus cluster if the value (i, j) on the consensus matrix is greater than 0.5. It should be mentioned that this ensemble clustering algorithm can result in a greater number of cluster than K, especially when the clustering methods give quite different results.

## 2. Regression Process

The aim of this step is to fit KPI values from selected network features. At this point, computations are done for each choice K of clusters' number, each value v of FailEstab and each cluster i. Values of cells belonging to the same cluster are aggregated. Then, different regression algorithms of the KPI feature as a function of network resource features are executed. Those algorithms are computed after taking the logarithm, on the training set. They are defined to handle linear and non-linear relationships between KPI and network resource features. Five of them are used: the linear regression, a generalized additive model (GAM, following Wood (2000) and Hastie and Tibshirani (1990)), a gradient boost method (GBM, following Burges (2010), a single-hidden-layer neural network (see Ripley (1996) for reference) and the multivariate adaptive regression splines method (MARS, see Friedman (1991)).

After obtaining predictions, the best regression algorithm is deduced for each cluster. For this purpose, the error rate (ER) shown on (2) is computed. On (2), $\left(y_l\right)$ stands for KPI data and $\left(\hat{y}_l\right)$ for fitted values, obtained on the cross-validation set.

$$ER\left(\left(y_l\right),\left(\hat{y}_l\right)\right) := \sum_l \left|y_l - \hat{y}_l\right| \tag{2}$$

Two quantities are used to deduce the best number of clusters K. The first one is ER(i, v, K), computed for the regression giving the lowest ER in (2). It measures error of fit. The other quantity is called cluster separation. It measures how distant are clusters each other, by checking if considering different clusters give better results instead of taking only one cluster. In other words, it sees if each cluster describes a specific consumer behavior. This quantity is defined in (3). To compute it, for all clusters $i, j \in \left\{1,\ldots,K\right\}$, $\left(y_l^i\right)$ are KPI values for the cross-validation set for cluster i, and $\left(\hat{y}_l^{i,j}\right)$ are fitted values of $\left(y_l^i\right)$ using the best prediction function obtained for cluster j.

$$Sep\left(i, v, K\right) := Sep\left(\left(y_l^i\right)\right) := \left(K-1\right)\frac{\sum_l\left|y_l^i - \hat{y}_l^{i,i}\right|}{\sum_{j\neq i}\sum_l\left|y_l^i - \hat{y}_l^{i,j}\right|} \tag{3}$$

To understand (3), one can look at the case where the prediction is not dependent of the cluster. Then, terms in each sum of both numerator and denominator are equal, and lead to a final value of 1. If however all predictions are well fitted and lead to different predictions, we can expect to have Sep(i, v, K) smaller than 1 for all clusters i.

To combine the different clusters and FailEstab values, those two quantities are weighted. Clusters are weighted with w(i | v, K), the proportion of values in cluster i among all clusters when FailEstab is v. FailEstab values are weighted with w(v) the proportion of values with FailEstab = v. Two summarizing values are deduced for each number of cluster K: one related to ER and the other to cluster separation. They are computed similarly, as shown on (4).

$$F\left(K\right) = \sum_v\sum_i F\left(i, v, K\right)w(i \mid v, K)w\left(v\right), F \in \left\{ER, Sep\right\} \tag{4}$$

Finally, the selection of the best cluster K is deduced with a combination of ER values and Sep values. This summarizing value is computed following (5). Number of clusters K with the lowest Summ value is selected.

$$Summ\big(K\big) = \log\big(ER\big(K\big)\big) + \log\big(Sep\big(K\big)\big) \tag{5}$$

This paragraph explains the choice to consider Summ value. If a small number of clusters K is selected, a large amount of data is available for each cluster but the clusters are not specific and can lead to an high bias in predictions. Therefore, ER and Sep values are first expected to decrease when increasing the number of cluster K (while the data training set is sufficiently large). But if a larger number of clusters K is selected, each cluster is specific to a behavior but few data are available and predictions suffer a high variance. Then, ER will begin to increase and in addition, the separation values will also increase when there is no more separation between clusters. On the whole, to choose K with the lowest Summ value is a compromise between those two extremes.

The last step is to retrieve the best regression predictions for this K, for each value of FailEstab and each cluster.

## 3. Deduction of a Comfortable Zone

For each value of FailEstab and each cluster, a comfortable zone is deduced. From prediction functions obtained with the selected number of clusters K, this zone is where KPI is greater than a fixed value, for example 99%. In this zone, accessibility will remain high. Therefore, for a new record, we will be able to predict if this record takes place in the low performance or the high performance zone.

## RESULTS

The global methodology is applied on collected data introduced in Section II. In the following, the studied KPI is the RAB Setup Success Ratio of all PS services (written VS.PS.RAB.Setup.Succ.Ration.CELL.custom.cell in plots). From resources features, four of them have been selected: number of HSUPA (High-Speed Uplink Packet Access) RAB Establishment Attempts in a cell (written VS.HSUPA.RAB. AttEstab), average Number of HSUPA UEs (user equipment) in a cell, average of upload link connection credit used and average of download link connection credit used.

In the following, ensemble clustering is first applied. Then, selection of the number of clusters K is obtained from ER and separation computations. Finally, comfortable zones are deduced from prediction functions.

## 1. Clustering Accuracy

Following our methodology, an ensemble clustering is obtained for each K from 1 to 20. An example of clustering visualization is shown on Figure 3 for K = 7, clusters 3 and 5 and for the selected feature VS.HSUPA.RAB.AttEstab. On this figure, data are restricted to FailEstab = 2 and plots are shown after taking logarithm. A good separation between clusters is observed, with a specific behavior in each cluster. On Figure 3 (a), the points are constrained into a linear zone, and the residual variance is almost homogeneous while network resource varies. On Figure 3 (b), the KPI is not a linear function of the network resource. Furthermore, points are noisy, especially for low values of the network resource. This latter behavior is due to inner variance in some cells rather than clustering performance, and is difficult to reduce here.

## 2. Number of Clusters Selection

Afterwards, the different regressions are computed. Error rate (ER) and clustering separation (Sep) are retrieved for each K, leading to the summarizing value (Summ). On Figure 4, the evolution of those three values is plotted as a function of the number of clusters K. On Figure 4 (a), the ER has a general decreasing trend while the number of clusters increases. The trend is also fluctuating. The lowest ER value corresponds to the highest number of clusters. On Figure 4 (b), the Sep value first follows a decreasing trend and then stabilizes with a slight increase (as expected in Section IV-B). The trend is similar on Figure 4 (c), showing more pronounced fluctuations. Taking the minimal Summ value from this figure, choice of K = 7 is done.

## 3. Comfortable Zone Identification

Finally, after choosing K = 7, regression predictions are used to fit data and deduce a comfortable zone for each cluster and each value of FailEstab. On Figure 5, obtained zone is plotted for two clusters, FailEstab = 2 and feature VS.HSUPA.RAB.AttEstab. On Figure 5 (a), the KPI steadily increases while the number of establishment attempts grows, to finally reach 99% value. Points related to each cell of the cluster have a limited variance, and the fitted line follows the scatter plot. On Figure 5 (b), accessibility reaches 99% more quickly on the fitted line. Then, a lower value of the network feature is necessary to reach accessibility. Contrary to the previous figure, some cells of this cluster have higher residuals around the fitted line.

*Figure 3. From data set, ensemble clustering obtained with K = 7, FailEstab = 2, for two clusters i, after taking logarithm. The plot represents aggregated cells with KPI as a function of a network resource. (a) i = 3; (b) i = 5*

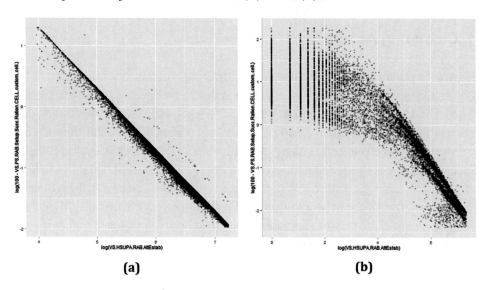

(a)            (b)

*Figure 4. From data set, evolution of a value as a function of number of clusters K. The red point indicates the minimal value. From (c), the selected K is 7. (a) ER value; (b) Sep value; (c) Summ value*

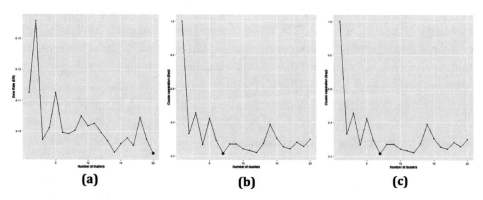

(a)        (b)        (c)

## DISCUSSION

The main implication of the work is to help mobile operators to elaborate a strategy of development of their network. They can have objective to achieve about accessibility, for example to reach 99% for some KPIs. Therefore, they have to determine where they should invest precisely and to elaborate network planning. Results de-

*Figure 5. From data set, KPI as a function of a network resource for K = 7, FailEstab = 2 and two clusters i. The blue line corresponds to the best regression fit for the cluster. The square is the 99% limit for accessibility evaluation. On the bottom, a lack of accessibility is shown in red and the comfortable zone in green, as a function of the current selected feature. (a) i = 3; (b) i = 5*

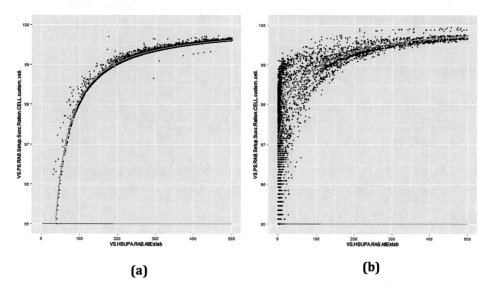

**(a)**                    **(b)**

duced with the presented methodology can point out which network features mobile operators should increase to improve network accessibility. Also, the amount of resource can be quantified and is specific to each cluster. For comparison, on Figure 2 (a) and Figure 2 (c), to get a 99% accessibility value is difficult, whereas on Figure 5, a specific and accurate value is obtained. Consequently, mobile operators can optimize resources to allocate for each type of cell. In this way, efficient and customized networks may be built.

## CONCLUSION

A new wireless analytic methodology to evaluate wireless network accessibility performance is performed in this paper. The purpose is to detect when the network starts to become less accessible. The methodology is based on relationship between accessibility and network resource measurements. The non-homogeneous behavior of cells of the network is taken into consideration by clustering them into related sets. Also, the procedure is automatic and focus on robust computations. Finally, comfortable zones are identified and indicate when accessibility begins to degrade.

A numerical study is performed to support our methodology, using data coming from China Mobile Communications Corporation.

The clustering processing scheme is a promising approach to improve accessibility predictions of wireless networks, paired with the combination of different regressions. Thereafter, new insights could be carried out to understand "in cell" variance, which remains important for some cells.

## REFERENCES

Amzallag, D., Bar-Yehuda, R., Raz, D., & Scalosub, G. (2013). Cell selection in 4G cellular networks. *IEEE Transactions on Mobile Computing*, *12*(7), 1443–1455. doi:10.1109/TMC.2012.83

Bezdek, J. C. (2013). *Pattern recognition with fuzzy objective function algorithms.* Berlin, Germany: Springer Science & Business Media.

Burges, C. J. C. (2010). From ranknet to lambdarank to lambdamart: An overview. *Learning*, *11*, 23–581.

Engels, A., Reyer, M., Xu, X., Mathar, R., Zhang, J., & Zhuang, H. (2013). Autonomous self-optimization of coverage and capacity in LTE cellular networks. *IEEE Transactions on Vehicular Technology*, *62*(5), 1989–2004. doi:10.1109/TVT.2013.2256441

Fred, A. (2001). Finding consistent clusters in data partitions. In *MCS'01 Proceedings of the Second International Workshop on Multiple classifier systems* (pp. 309-318). London, UK: Springer. doi:10.1007/3-540-48219-9_31

Friedman, J. H. (1991). Multivariate adaptive regression splines. *Annals of Statistics*, *19*(1), 1–67. doi:10.1214/aos/1176347963

Hastie, T. J., & Tibshirani, R. J. (1990). *Generalized additive models* (Vol. 43). CRC Press.

Hu, S., Ouyang, Y., Yao, Y., Fallah, M. H., & Lu, W. (2014). A study of LTE network performance based on data analytics and statistical modeling. In *2014 23rd Wireless and Optical Communication Conference (WOCC)* (pp. 1-6). IEEE.

Kaufman, L., & Rousseeuw, P. (1987). *Clustering by means of medoids.* Delft, The Netherlands: North-Holland.

Kaufman, L., & Rousseeuw, P. J. (2009). *Finding groups in data: an introduction to cluster analysis* (Vol. 344). New York: John Wiley & Sons.

Khan, A., Sun, L., & Ifeachor, E. (2012). QoE prediction model and its application in video quality adaptation over UMTS networks. *IEEE Transactions on Multimedia*, *14*(2), 431–442. doi:10.1109/TMM.2011.2176324

MacQueen, J. (1967). Some methods for classification and analysis of multivariate observations. In *Proceedings of the fifth Berkeley symposium on mathematical statistics and probability* (Vol. 1, pp. 281-297). Oakland, CA: University of California Press.

McLachlan, G. J., & Basford, K. E. (1988). *Mixture models: Inference and applications to clustering*. New York, US: Dekker.

Miller, A. (2002). *Subset selection in regression*. CRC Press. doi:10.1201/9781420035933

Navaie, K., & Sharafat, A. R. (2003). A framework for UMTS air interface analysis. *Canadian Journal of Electrical and Computer Engineering*, *28*(3/4), 113–129. doi:10.1109/CJECE.2003.1425098

Ouyang, Y., & Fallah, M. H. (2010). A performance analysis for UMTS packet switched network based on multivariate KPIs. In *Wireless Telecommunications Symposium (WTS)*, 2010 (pp. 1-10). IEEE. doi:10.1109/WTS.2010.5479629

Ouyang, Y., Fallah, M. H., Hu, S., Yong, Y. R., Hu, Y., Lai, Z., & Lu, W. D. et al. (2014). A novel methodology of data analytics and modeling to evaluate LTE network performance. In *Wireless Telecommunications Symposium (WTS)*, 2014 (pp. 1-10). IEEE.

Ouyang, Y., & Yan, T. (2015). Profiling wireless resource usage for mobile apps via crowdsourcing-based network analytics. *IEEE Internet of Things Journal*, *2*(5), 391–398. doi:10.1109/JIOT.2015.2415522

Ouyang, Y., Yan, T., & Wang, G. (2015). CrowdMi: Scalable and diagnosable mobile voice quality assessment through wireless analytics. *IEEE Internet of Things Journal*, *2*(4), 287–294. doi:10.1109/JIOT.2014.2387771

Ripley, B. D. (1996). *Pattern recognition and neural networks*. Cambridge, UK: Cambridge University Press. doi:10.1017/CBO9780511812651

Szlovencsak, A., Godor, I., Harmatos, J., & Cinkler, T. (2002). Planning reliable UMTS terrestrial access networks. *Communications Magazine, IEEE*, *40*(1), 66–72. doi:10.1109/35.978051

Tsao, S., & Lin, C. (2002). Design and evaluation of UMTS-WLAN interworking strategies. In *Vehicular Technology Conference, 2002. Proceedings. VTC 2002-Fall. 2002 IEEE 56th* (Vol. 2, pp. 777-781). IEEE. doi:10.1109/VETECF.2002.1040705

Wood, S. N. (2000). Modelling and smoothing parameter estimation with multiple quadratic penalties. *Journal of the Royal Statistical Society. Series B, Statistical Methodology*, *62*(2), 413–428. doi:10.1111/1467-9868.00240

# Chapter 3
# Modeling for Time Generating Network:
## An Advanced Bayesian Model

**Yirui Hu**
*Rutgers University, USA*

## ABSTRACT

*Modeling co-occurrence data generated by more than one processes in network is a fundamental problem in anomaly detection. Co-occurrence data are joint occurrences of pairs of elementary observations from two sets: traffic data in one set are associated with the generating entities (Time) in the other set. Clustering algorithms are valuable because they can obtain the insights from the varied distribution associated with generating entities. This chapter leverages co-occurrence data that combine traffic data with time, and compares Gaussian probabilistic latent semantic analysis (GPLSA) model to a Gaussian Mixture Model (GMM) using temporal network data. Experimental results support that GPLSA holds better promise in early detection and low false alarm rate with low complexity of implementation in a fully automatic, data-driven solution.*

## INTRODUCTION

Network monitoring devices become more powerful in recent years, which can collect data at high rates. We can then extract relevant information from a large amount of noisy data and design an effective anomaly detection system (Ahmed, Oreshkin

DOI: 10.4018/978-1-5225-1750-4.ch003

& Coates, 2007). In many applications, the collected data are generated by more than one processes, namely co-occurrence data (Aggarwal, 2013). Co-occurrence data are joint occurrences of pairs of elementary observations from two sets: traffic data (Observations-*W*) in one set are associated with the generating entities (Time stamp or node ID-*D*) in the other set. Modeling co-occurrence data (traffic data with generating entity) is a fundamental problem in anomaly detection. It poses a challenge for effective anomaly detection when the usual distribution varies with generating entities.

This chapter focuses on temporal network data generated with generating entities, and discusses several unsupervised multiple cluster-based analytical models in anomaly detection. We first introduce several candidate models including GMM and presents GPLSA model using Expectation Maximization algorithm, then apply "Donut" algorithm for anomaly detection and compare the detection rates between GMM and GPLSA with some examples.

## BACKGROUND

As we discussed in previous chapter, the optimal choice of model usually comes with a good understanding of domain knowledge. Cluster-based analytical models are able to identify the underlying similar patterns for temporal traffic data.

A number of studies have used GMM for detection problems as described in Tax and Duin (1998) and Desforges, Jacob and Cooper (1998). One can develop an overall model that looks at the entire data collected over all time slots without explicit regard to the time variable. GMM is a probabilistic clustering algorithm that assumes all the data points are generated from a mixture of finite number Gaussian distribution, whose parameters are estimated using EM algorithm. Mathematics form of GMM is derived in previous chapter.

However, in most applications for co-occurrence data where observations are generated by time or node ID, the usual distribution often varies with generating entities, which poses a challenge for effective anomaly detection. GMM has advantage when clusters have different sizes and correlations within them, while ignoring the generating entities may result in a missed detection in co-occurrence data. For example, if we evaluate the likelihood of a testing data point, the clustering model may yield a result of 'likely', whereas in reality it is unlikely for the underlying time.

Our work thus extends GMM to a Bayesian probabilistic solution through a Bayesian model. We implement GMM as a reference model, and compare to a Bayesian probabilistic model called GPLSA.

*Figure 1. Matrix decomposition of PLSA*

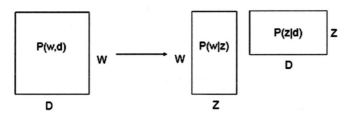

## Issues, Controversies, Problems

The challenge is to learn model/distribution which can explicitly explain the relationship between *X* and its generating entities *G* using clustering techniques. Bayesian framework models organize co-occurrence data into multiple clusters by identifying latent mapping instead of generating entities. Having learned such models, we can then automatically characterize the nature of normal traffic behaviors and detect abnormal patterns with due respect to the underlying the time generating entities.

Probabilistic latent semantic analysis (PLSA) is a statistical model for the analysis of co-occurrence data, which maps co-occurrence data to a latent variable. The core of PLSA is to discover the underlying semantic structure of co-occurrence data under a probabilistic framework. It was first developed by *T*. Hoffman (1999a) and it was initially used in text mining, however, its use spread shortly in other 'prediction centric' applications of machine learning such as recommender systems. We believe that this is the first instance where PLSA has been applied for anomaly detection.

The technique of PLSA is first used in text community, which is very similar to the singular value decomposition (Figure 1). PLSA expresses our data in terms of 3 sets:

- **W:** Observed traffic data *w* (a vector of p dimensions) in $W=\{ w_1, ..., w_N \}$. Let *N* be the number of data.
- **D:** Observed generating entity *d* in $D=\{ d_1, ..., d_t \}$. Let *d* be the number of discrete values in *D*.
- **Z:** Latent variables *z* in $Z=\{ z_1, ..., z_k \}$. The number of latent variables is specified a priori.

The goal of PLSA is to extract the so called "latent variables" and explain the generating entity as a mixture of them. Figure 2 illustrates the intermediate layer of latent variables that link the traffic data and the generating entity (Hofmann, 1999b). Each generating entity can be explained as a mixture of latent variables weighted with probability $P(z|d)$ and each traffic data expresses a latent variable with probability $P(w|z)$. In particular, Gaussian PLSA (GPLSA) assumes $w|z$ to be normally distributed.

## SOLUTIONS AND RECOMMENDATIONS

The solution is based upon a Bayesian probabilistic model – GPLSA holds better promise to accurately consider the generating entities with more sensitivity (early detection) and relatively low complexity. In particular, Gaussian Probabilistic Latent Semantic Analysis (GPLSA) is a natural extension of GMM through a Bayesian model.

The key assumption of PLSA is the conditional independence which means that traffic data and generating entity are conditionally independent given the latent variable. Using the conditional independence assumption, the model can be defined by the joint distribution.

$$P\left(w|d\right) = \sum_{z} P(w \mid z) P(z \mid d),\tag{1}$$

$$P\left(\mathrm{w},\ \mathrm{d}\right) = \sum_{z} P\left(z\right) P(d \mid z) P(w \mid z),\tag{2}$$

*Figure 2. The general structure of PLSA model*

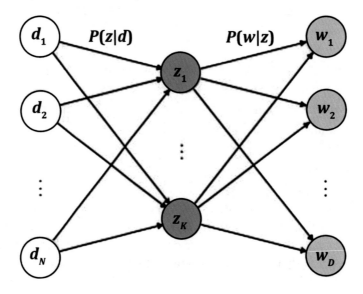

### Modeling for Time Generating Network

Another assumption is the joint variable $(d, w)$ is independent sampled, and consequently, the joint distribution of the observed data can be factorized as a product.

We then calculate the log-likelihood function and derive the maximum likelihood estimate (MLE) using EM algorithm. Standard calculation yields the following steps:

E-step:

$$P\left(z|d,w\right) = \frac{P\left(z\right)P(d\mid z)P(w\mid z)}{\sum_{z'\in Z}P\left(z'\right)P(d\mid z')P(w\mid z')}, \tag{3}$$

M-step:

$$P\left(w|z\right) \propto \sum_{d} n\left(d,w\right)P(z\mid d,w), \tag{4}$$

$$P\left(d|z\right) \propto \sum_{w} n\left(d,w\right)P(z\mid d,w), \tag{5}$$

$$P\left(z\right) \propto \sum_{d}\sum_{w} n\left(d,w\right)P(z\mid d,w). \tag{6}$$

In GPLSA, we assume that $P(w|z)$ follows MVG, with $\theta = (\mu,\Sigma)$. The EM steps can be rewritten as follows:

E-step:

$$P\left(z|d,w\right) = \frac{P(w\mid z)P(z\mid d)}{\sum_{z'\in Z}P(w\mid z')P(z'\mid d)}, \tag{7}$$

M-step:

$$P\left(z|d\right) = \frac{\sum_{d'=d}P(z\mid d,w)}{\sum_{d}\sum_{d'=d}P(z\mid d,w)}. \tag{8}$$

Similarly, we obtain the parameters of Gaussian distributions:

$$\mu_{d,w} = \frac{\sum_{w=w'} P(z \mid d, w)w}{\sum_{w=w'} P(z \mid d, w)}, \tag{9}$$

$$\sum_{d,w} = \frac{\sum_{w=w'} P(z \mid d, w)\left(w - \mu_{d,w}\right)'\left(w - \mu_{d,w}\right)}{\sum_{w=w'} P(z \mid d, w)}. \tag{10}$$

## EXPERIMENTAL RESULTS

To advocate the applicability of machine learning algorithms to network anomaly detection, we collect measured traffic data at Radio Network Controller (RNC) level in the wireless networks in a given city. Data are collected from a bunch of RNCs every 30 minutes for 90 days. We then separate the dataset into train (first 80 days) and test data (last 10 days).

In traffic data, there are many key performance indicators (KPIs). We pick the following two KPIs because their behaviors follow a 24-hour cycle with regular patterns. CS_ErLang (CS indicates circuit switch) measures the number of calls and PS_Throughput (PS represents packet switch) measures the data flow.

In co-occurrence data *(D, W)*, *W* represents 2-dimensional traffic vector of CS and PS. *D* stands for the generating entity - time, which is a discrete variable with values 1, ..., 48.

We first train GPLSA for *K=8* and estimate model parameters using EM. The parameters include Gaussian distribution with mean ($\mu_1, ..., \mu_k$) and covariance matrix $\left(\sum_1, ..., \sum_k\right)$ for each latent state; the multinomial distribution of latent states with conditional probability P(z|d) given a specific time d. The GPLSA model we learned can also be described in the following generative process:

For each co-occurrence pair *(d, w)*, estimate the probability *P(d)* of generating entity (time slot). For each observation $w_i, i = 1, ..., N$, at time $d, d = 1, ..., 48$. Select a latent state from a multinomial conditioned at given time *d* with probability *P(z|d)*. Select a traffic vector $w_i$ from a learned Gaussian distribution on the previous chosen latent state.

Figure 3 can be first visualized as scatter plot for train data with CS and PS. More importantly, it is an overview of GPLSA mapping: from co-occurrence data

*Figure 3. Overview of mapping co-occurrence data to GPLSA latent states*

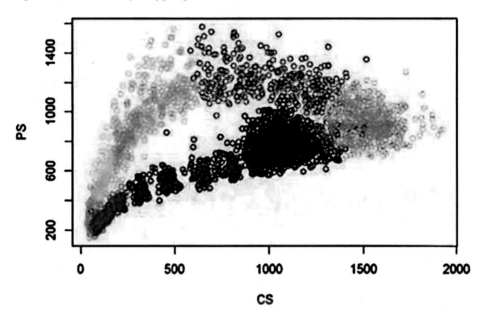

*(D, W)* to latent states *(Z)*. The 8 colored latent states then link the traffic data (CS and PS) and the time using the above probabilistic model.

Note here that GMM and GPLSA share similar Gaussian clusters with mean and covariance matrix. The major difference is: *P(z|d)* is different given different time slot in GPLSA; *P(z|d)=P(z)* in GMM.

Having learned the machine learning models after filtering, we are able to detect anomaly in test data. There are a bunch of ways to define anomaly using different statistics and tests. In this paper, we introduce a novel "Donut" algorithm for detect abnormal patterns in test data.

The advantage of "Donut" algorithm is that it can automatically detect two types of anomaly: Type 1 anomalies are single traffic data points with extreme unlikely log-likelihood given trained model; Type 2 anomalies are unlikely traffic data series whose log-likelihoods persist within the unlikely thresholds in a certain length of window.

Figure 4 illustrates the 3 zones defined by "Donut" algorithm using GMM and GPLSA. The red and green line represents 1% and 10% quantile.

RNC level data is considered to be very stable without much anomaly, thus it has limitation in anomaly detection. Although our experiments are performed on a limited data set, the log-likelihood results can still provide insights in model selection:

*Figure 4. Illustration of three zones using GMM and GPLSA in test data after filtering*

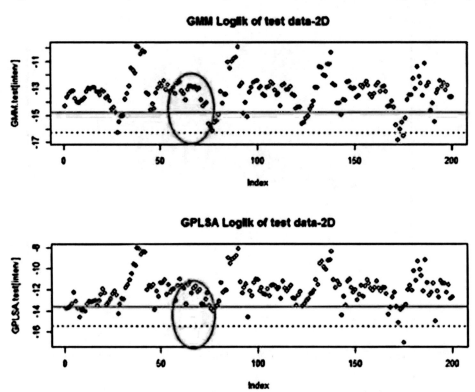

From the standpoint of log-likelihood, GPLSA has greater averaged log-likelihood value than GMM. Log-likelihood measures how likely the data are given selected models, which support GPLSA can better explain the data. From the standpoint of false alarm rate, GPLSA report less anomalies than GMM. Given the "pure" data we filtered, we believe that GPLSA is the model of choice in anomaly detection.

## FUTURE RESEARCH DIRECTIONS

Data that is continuously being generated from machines, mobile devices, and network data traffic contains valuable information into business operations (Aggarwal, 2013). After detecting anomalies, the next step is to perform causality analysis for a close loop corrective actions as well as preventing future situations

Future network analysis is to develop a self-diagnosing system that triggers self-optimizing and self-healing wherein anomaly detection plays a critical role. It is hence clear that anomaly detection must be accurate and sensitive.

## CONCLUSION

Several clustering-based unsupervised machine learning algorithms for traffic data with generating entities are discussed in this paper. Experimental results performed on RNC traffic data indicate that GPLSA holds better promise in anomaly detection with regards to both earlier detection and lower false alarm rate.

## REFERENCES

Aggarwal, C. C. (2013). *Outlier Analysis*. New York: Springer. doi:10.1007/978-1-4614-6396-2

Ahmed, T., Oreshkin, B., & Coates, M. J. (2007). Machine learning approaches to network anomaly detection. In *SYSML'07 Proceedings of the 2nd USENIX workshop on Tackling computer systems problems with machine learning techniques*. Berkeley, CA: USENIX Association.

Desforges, M. J., Jacob, P. J., & Cooper, J. E. (1998). Applications of probability density estimation to the detection of abnormal conditions in engineering. *Proceedings - Institution of Mechanical Engineers*, *212*(8), 687–703.

Hofmann, T. (1999a). Probabilistic latent semantic indexing. In *Proceedings of the 22nd annual international ACM SIGIR conference on Research and development in information retrieval (SIGIR '99)* (pp. 50-54). New York, NY: ACM. doi:10.1145/312624.312649

Hofmann, T. (1999b). Probabilistic latent semantic analysis. In *Proceedings of the Fifteenth conference on Uncertainty in artificial intelligence* (pp. 289-296). Burlington, MA: Morgan Kaufmann Publishers Inc.

Tax, D. M. J., & Duin, R. P. W. (1998). Outlier detection using classier instability. In *Proceedings of the Joint IAPR International Workshops on Advances in Pattern Recognition* (pp. 593-601). London, UK: Springer. doi:10.1007/BFb0033283

## KEY TERMS AND DEFINITIONS

**Anomaly Detection:** The identification of items, events or observations which do not conform to an expected pattern or other items in a dataset.

**Appalachia:** A geographic and cultural region of the Mideastern United States. The population in media is portrayed as suspicious, backward, and isolated.

**Causal Analysis:** Analysis that aims at identify and understand the reasons why things are as they are and hence enabling focus of change activity.

**Ethnocentric:** A belief that one's own culture is superior to other cultures.

**Family-Centricity:** The belief that family is central to well being and that family members and family issues take precedence over other aspects of life.

# Chapter 4

# Identifying Dissatisfied 4G Customers from Network Indicators:
## A Comparison Between Complaint and Survey Data

**Xinling Dai**
*Nanjing Howso Technology, China*

## ABSTRACT

*Feedback data directly collected from users are a great source of information for telecom operators. They are usually retrieved as complaints and survey data. For the mobile telecoms sector, one purpose is to manage those data to identify network problems leading to customer dissatisfaction. In this paper, a quantitative methodology is used to predict dissatisfied users. It focuses on extraction and selection of predictive features, followed by a classification model. Two sets of data are used for experiments: one is related to complaints, the other to survey data. Since the methodology is similar for those two sets, prediction efficiency and influence of features are compared. Specific influence of user loyalty in survey data is highlighted. Thus, the methodology presented in this article provides a reference for the mobile operators to improve procedures for collecting feedback answers.*

DOI: 10.4018/978-1-5225-1750-4.ch004

# INTRODUCTION

With development of high speed LTE connections, a major increase of network consumption occurred. Telecom operators have to identify network issues. It is an important task, especially when accessibility of network deteriorates or when some services become unreachable. A great source of information to understand those problems is to collect data directly from customers. Those data allow to get a feedback from customers and are commonly of two types: surveys and complaints data. User surveys are an active way to retrieve customers' appreciation of a service. Complaints are a passive way to obtain information, underlining the service problems. Surveys give a diffuse appreciation over time, whereas complaints data are focused on a specific issue. Those data are mostly subjective, costly to produce (for surveys) or in a limited number (for complaints). Furthermore, a latency period is observed between emergence of a problem and its report to mobile operators.

This article specifically focuses on issues related to 4G networks. Network key performance indicators – hereinafter called features – are used in addition with feedback data to identify dissatisfied 4G customers and origin of dissatisfaction. Those features represent a continuous and huge flow of data, directly collected by mobile operators. Therefore, the main task is to link feedback data with network features. The methodology presented in this article selects informative features in an iterative process by using machine learning tools. Then, qualitative information can be retrieved from feedback data, allowing to understand origin of dissatisfied customers and to provide customer satisfaction predictions. An important characteristic is to perform an identical process for both survey and complaints data. As the source of data is similar, the information they contain can be compared. The discussion focuses on two aspects: first, the predictive quality of the information contained in those data; then, the difference in the origin of dissatisfaction. The main result is to observe an external factor (user loyalty to mobile operator) interfering survey effectiveness. Consequently, this methodology provides a guidance for the mobile operators to improve survey question quality and effectiveness.

The article is organized as follows. In the second section, related works are presented, and last paragraph compares explicit innovative results to existing works; in the third section, data retrieved and needed for computations are presented; In the fourth section, the modeling and algorithmic process are presented; results are then shown in fifth section; finally, conclusion and further improvements are given in the final section.

## RESEARCH BACKGROUND

The customer satisfaction evaluation through customer data has been an extended field of research for industrial companies. The concept of customer loyalty has been intensively studied in Alan and Basu (1994), and Olever (1999), as well as the customers' general behavior (Olever, 1997). From those concepts, the idea of a unique Net Promoter Score (NPS) emerged (Reichheld, 2003), which is a one-question survey focusing on how likely a customer would recommend the service to other people. From a more general perspective, how to carry out a survey process has been summarized. A review is available in Fowler (2013), describing how to conduct a survey, to design questions and to sample users.

Common techniques to analyze survey data is presented in Rossi, Wright and Anderson (2013). In the telecom industries, analysis of survey data became a key research subject to provide better services for customers. In Qian (2011), and Rossi et al. (2013), links between surveys and external indicators have been studied. Also, Yu (2014) brings an analysis on network quality satisfaction through a survey, determining which main factors influence the network quality.

As for complaints analysis, an overview of available information from those data can be seen in Goodman and Newman (2003). This article points out the root cause of a complaint generally cannot be directly deduced. Prediction tasks have also been performed using complaints data, including big data techniques (Hadden, Tiwari, Roy, & Ruta, 2006).

The approach in this article is different because an identical methodology is used both for survey and complaints data, leading to a quantitative analysis based on network data. Explicitly, described results allow to: deduce network features impacting the emergence of a complaint or of a low survey score; compare reliability and efficiency between survey and complaints data; give insights to explain why survey reliability is lower, to identify source of noise and to deduce initial survey design drawbacks.

## COLLECTED DATA

Two types of data are collected: complaints and survey data. They are both gathered from China Mobile Communications Corporation in Jiangsu province and anonymized.

Complaints data have been retrieved in October and November 2015. Only the 53 complaints which are related to network issues of 4G users have been selected. The set is completed with 368 users who did not send a complaint.

Survey data have been elaborated and answers collected in August and September 2015. 4013 answers have been retrieved. On Table 1, the different questions are listed. Only questions related to 4G network performance are kept, i.e. only items 1, 2, 3 and 6. The mean over those four answers is taken to get a 4G satisfaction score. To deduce a value sharing similarities with a complaint, the 20% worst scores are selected and related users are defined as unsatisfied. The 80% remaining users are defined as satisfied. This threshold has been selected following Pareto principle, by observing shape of scores' histogram. The separation corresponds to the score of 6 for this survey data set.

Those feedback sets are completed with general and network features. General indicators are for example age, gender or number of days since subscription. Network features correspond for example to webpage display success rate, time of HTTP response failure, number of wireless network redirections, etc. On the whole, there are about 80 different features. For network features, collection is performed at different dates prior to feedback, leading to different time series. Specifically, values are collected each day during the 30 days before survey answer, and each quarter of hour during the 2 hours before complaint.

Finally, two global data sets are obtained from same source of users and share a similar shape. For them, number of data collected is limited, being time consuming or costly to increase. Also, prediction objective is a single binary indicator (complaint/no complaint and unsatisfied/satisfied).

*Table 1. China Mobile's survey questions*

| Number | Question |
|--------|----------|
| 1 | Would you like to recommend China Mobile 4G to your friends or relatives? Please evaluate using an integer between 1 and 10. 1 means you're very reluctant to recommend it, 10 means you're very favorable. |
| 2 | Overall, how would you rate China Mobile's call quality? Please evaluate using an integer between 1 and 10. 1 means very bad, 10 means very good. |
| 3 | Overall, how would you rate China Mobile's 4G network quality? Please evaluate using an integer between 1 and 10. 1 means very bad, 10 means very good. |
| 4 | Overall, how would you rate flow package being used? Please evaluate using an integer between 1 and 10. 1 means very bad, 10 means very good. |
| 5 | Overall, how would you rate flow alert service of China Mobile? Please evaluate using an integer between 1 and 10. 1 means very bad, 10 means very good. |
| 6 | From your experience using China Mobile 4G, how would you rate overall performance? Please evaluate using an integer between 1 and 10. 1 means very bad, 10 means very good. |

# MODELING PROCESS AND ALGORITHM OVERVIEW

Process to perform the classification task is described in this section. The aim is to predict and separate unsatisfied users from non satisfied ones using general and network resource features. One main step is to prepare data before effectively applying a machine learning tool for classification. First, features are extracted and transformed from original data, to catch linear and non-linear behaviors. Then, a crucial step is to choose the predictive features. Selection is performed iteratively by maximizing a prediction score. The definition of this prediction score is presented, based on Receiver Operating Characteristic (ROC) curve computations. Finally, different machine learning tools tested for classification are depicted.

## Extraction and Transformation of Features

The initial step from a global data set is to extract and combine features explaining user dissatisfaction.

Network features take the form of short length numeric time series, as explained in "Collected Data" section. Two informative features are first extracted from those time series: mean and standard deviation over period.

The second step is to perform transformation over all numeric features. In fact, many indicators have values near 0 (for example counting a number of occurrences, or a percentage of failure) or 100 (for example a success rate shown in percentage). Also, empirical distribution often displays a heavy-tailed behavior. Consequently, operations are applied over features. Depending of features, those operations are natural logarithm, $\log(1 + .)$ or $\log(100 - .)$. Original features are also kept. Box-Cox power transformations (Guerrero & Johnson, 1982; Sakia, 1992) have also been tested, where the parameter has been fit to have the highest correlation with feedback value. Those latter transformations have not been selected finally to prevent over fitting.

The third step is specific to categorical features. For each categorical indicator, the 4 most common levels are kept, whereas other ones are let in a single group "others".

Consequently, a large quantity of indicators are extracted and has to be managed. As sample size is small, an automatic process is needed to perform a parsimonious selection of features.

## Algorithm to Select Features

The chosen iterative process to select features is close to the univariate filters method (Saeys, Inza & Larrañaga, 2007). Each feature is initially considered individually and a prediction score based on capacity for this feature to predict the feedback value is computed. Explanation of this prediction score is postponed in paragraph "Computation of Prediction Scores". Features are then ranked according to those scores and a percentage of features is kept, for example 40%. Afterwards, all combinations of two features among kept ones are used, each combination leading to a new prediction score. Again, 40% of features are kept and the iterative procedure continues. Loop is stopped when less than 10 features remain, and the 10 features leading to the best scores are selected. The formal algorithm can be summarized as follows:

```
Initially set kept_features as all features.
Initially consider combination with 1 element by setting: k = 1
While #kept_features > 10 {
    For each k-combination of kept features {
        Compute a prediction score
    }
        For each kept feature, take all k-combinations where
this feature appears and take mean over obtained prediction
scores.
        Select the p * #kept_features combinations with the best
scores as kept_features.
        k = k + 1
}
```

Prediction score is then computed on all 1024 subsets of the remaining 10 features to select the most suitable features. At this point, some of them can be manually removed based on correlation. By its iterative construction, the whole method is particularly suitable for high dimensional data. However, to make computations tractable, percentage of kept features should not be too high. For the collected feedback data, nbFeat = 310 and p = 0.4 are selected, leading to 310 computations of scores in the first step, 7626 (the number of 2-combination of a set of length nbFeat $\times$ p) in the second step, 18424 in the third step and 3876 in the fourth step.

## Computation of Prediction Scores

In this paragraph, it is assumed that some features have been selected. The purpose is to retrieve a unique score which assesses ability of those combined features to predict the feedback value. This prediction is developed to take into account specificity of the feedback value. As collected data sets are unbalanced, direct computation of accuracy is not suitable as a prediction score. Furthermore, permitted number of false error rate is not fixed and depends on unknown external factors (for example with complaints, a user which does not complaint can also have some 4G network difficulties). Thus, an approach based on ROC curves is privileged, rather than AIC or AICC (Bozdogan, 1987; Hurvich & Tsai, 1989) criterions. In an implementation point of view, those curves have been computed from classification scores with the ROCR package (Sing, Sander, Beerenwinkel, & Lengauer, 2005) of R.

Chosen measure is the signed distance from ROC curves to the identity function. This prediction score shares some similarities with Area Under Curve (AUC) (Cortes & Mohri, 2004; Huang & Ling, 2005). However, the AUC method has drawbacks, mostly because it performs mean over all false positive rate with different weights, making difficult a clear model comparison (Hand, 2009; Hand & Anagnostopoulos, 2013; Lobo, Jiménez-Valverde & Real, 2008). By contrast, the chosen measure score does not consider general aspect of the ROC curve but only a specific part.

It remains to discuss how ROC curves are obtained. Those are based on different classification models such as logistic regression or random forest. Those models are discussed in paragraph "Machine Learning Algorithms for Classification".

## Machine Learning Algorithms for Classification

To select and classify unsatisfied vs. satisfied users, a classification algorithm task is performed from selected features. Three different algorithms are tested, listed thereafter.

Logistic regression (Hosmer & Lemeshow, 2004) is a regression model usually dedicated to binary classification. This is a classic algorithm, included in the generalized linear model. As a linear classifier, it cannot catch complex non-linear patterns. Also, it is sensitive to correlation within predictive variables. Thus, correlations have to be checked to avoid over-fitting and over-confidence in certain variables. Its main advantage is to be quick to run and reliable for linear patterns identification. Furthermore, it is a white box model, where impacting features are easy to understand.

Random Forest (RF) (Liaw & Wiener, 2002) is an ensemble learning method (Dietterich, 2000) which can be used for binary classification, among others. It is based on construction of a great quantity of decision trees, each tree being computed without considering earlier trees. Those trees are then combined to obtain a single prediction. This approach, called bagged trees, have several advantages: First, it is less subject to over fitting than for example boosted trees methods; Then, it is a white box model, where importance of each feature can be understood. Trees are also relatively fast to train.

Support Vector Machine (SVM) (Suykens & Vandewalle, 1999) is a supervised learning model usually used for binary classification. It is a geometric method, since the classic algorithm tries to separate feature space into two hyperplanes, by minimizing empirical classification error and maximizing geometric margin. Also, fitting the regularization parameter allows to get soft margins. This method was initially developed as a linear classifier, but the so-called "kernel trick" allows to perform non-linear classifications.

In an implementation point of view, those algorithms are computed with caret package (Kuhn, 2008) of R. For each method, k-fold cross validation is performed. Once the final features have been selected, the related classification model is set and chosen, from which ROC curves are deduced.

## RESULTS

In order to evaluate the methodology presented in this article, the two feedback data sets described in "Collected Data" section are used: complaints data and survey data. Features are extracted according to "Modeling process and algorithm overview" section and prediction models are built. First, quantitative prediction results are shown for each set and accuracies are compared. Then, a qualitative description of selected features' impact is done. Again, complaints and survey data are compared. Finally, results are discussed to understand observed differences. Some insights concerning improvements of the methodology are proposed.

### Quantitative Prediction Results

For the two data sets, computations are performed with logistic regression, RF and SVM. For complaints data, percentage of selected features at each iterative selection step is set to p = 40% and number of k-fold is 4. For survey data, p = 40% and k-fold = 10. A lower number of k-fold was selected for complaints data because of the small size of the set (separating into 10 classes let a high probability to have a class without any complaint). Finally, logistic regression algorithm is retained, producing the best accuracy results.

To show prediction improvements compared to random predictions, results are summarized into a predictive table. Users are first separated into deciles and ranked by their scores (the higher the score, the higher the probability to be unsatisfied). Then, percentage of unsatisfied users in each decile over total number of unsatisfied users is deduced. Those values are obtained from ROC curves. Ideally, a greater percentage in the first deciles and a lower one in the last deciles are expected. If predictions had been performed randomly, each table cell should have been near 10%. Results are shown on Table 2, where the first column stands for complaints data and the second one for survey data.

For complaints data, more than 90% of observed unsatisfied users stand in the 4 first deciles. Therefore, a strong relationship exists between users experiencing a network issue and an effective problem in the network features.

*Table 2. Percentage of unsatisfied users in each decile, ranked by prediction scores*

| Ranked Users | Unsatisfied Users in the Decile over Total Unsatisfied Users (for Complaints Data) | Unsatisfied Users in the Decile over Total Unsatisfied Users (for Survey Data) |
|---|---|---|
| Top 10% | 48.6% | 18.6% |
| 10% to 20% | 15.1% | 16.4% |
| 20% to 30% | 15.0% | 14.7% |
| 30% to 40% | 11.6% | 12.4% |
| 40% to 50% | 0.0% | 5.1% |
| 50% to 60% | 1.7% | 5.0% |
| 60% to 70% | 0.0% | 9.1% |
| 70% to 80% | 3.5% | 6.1% |
| 80% to 90% | 0.0% | 8.0% |
| 90% to 100% | 4.5% | 4.6% |

For survey data, 62.1% of unsatisfied users stand in the 4 first deciles. Then, a relationship exists between user satisfaction and network features. However, about 40% of users with fair network indicators are not satisfied. Therefore, data coming from survey data are less predictive.

## Qualitative Impact of Features

For each data set, a predictive model has been computed and impact of each selected feature can be studied. Those features are divided into three different groups, for discussion purpose: group A is related to network indicators which have a positive impact on dissatisfaction; group B is related to network indicators which have a negative impact on dissatisfaction; group C includes general indicators which have a negative impact on dissatisfaction. Selected features are summarized in Table 3.

For complaints data, 5 features have been selected. In group A, there are "number of service request redirection", "log of the number of attach requests" and "TCP retransmission rate". In group B stand the remaining features, which are "log of the X2 handover success rate" and "standard deviation of TAU success rate". Impact of those features can be interpreted, except for the last one.

For survey data, 4 features have been selected. In group A, there are "log of number of service request redirection". In group B, "DNS query success rate" and "HTTP response delay". In group C, "number of days since subscription". Impact of those four features are understood.

*Table 3. Listing of selected features for each feedback data*

| Group | Complaints Data | Survey Data |
|---|---|---|
| A – network indicators with positive impact on dissatisfaction | • "Number of service request redirection"<br>• Log of the "number of attach requests"<br>• "TCP retransmission rate" | Log of number of "service request redirection" |
| B – network indicators with negative impact on dissatisfaction | • Log of the "X2 handover success rate"<br>• Standard deviation of the "TAU success rate" | • "DNS query success rate"<br>• "HTTP response delay" |
| C – general indicators with negative impact on dissatisfaction | | "Number of days since subscription" |

Selected features for complaint and survey data share similarities. For those two sets, group A and group B are present. Although features are different, they are highly correlated and represent a similar behavior. Also, since many couple of features have a high correlation, selected features cannot be called the "best ones" in a causal way, but only the most predictive. The main difference in feature selection is the presence of the group C for survey data, which is the number of days since subscription. This difference is discussed in the following section.

## Discussion of Results

Complaint results are more predictive compared to survey results. One cause is that complaints are an active way to reflect a problem, contrary to surveys. Then, complaints are directly related to a network problem whereas survey questions are less definite. Indeed, direct correlations between feedback score and the different features are low with the survey data, being less than 0.15 for all features. Inner variance is also important for the survey data, and external causes or subjectivity can affect answers.

Also, the survey questions are related to network satisfaction among others. In fact, a high correlation exists between the survey questions. For example, NPS (question 1 of Table 1) is positively correlated with network quality score (question 3 of Table 1), with a correlation about 70%. Therefore, users which do not recommend the service may grade network quality negatively, even if there is no network problem. To support this hypothesis, a specific impact of a feature related with loyalty is present in the survey model ("number of days since subscription").

Consequently, to remove general satisfaction influence in scores to keep network quality satisfaction may improve predictive results. One way can be to consider heterogeneity between survey answers, to discard users which answer a low score or a high score for all questions, and to only consider users which specifically scored low for the network quality question.

## CONCLUSION AND IMPROVEMENTS

A methodology to predict users' satisfaction is set up through an automatic process. This process includes feature extraction, feature selection by ranking scores of features and model for algorithm classification. Experiments are conducted on two sets, collected from different contexts: the first data set comprises users' complaints specific to 4G network quality issues, whereas the second set is related to survey answers. Following the same methodology for those two sets allows to compare prediction efficiency and qualitative features' influence. Results show that predictions obtained with survey data are less accurate than with complaints data, being less objective and subject to additional noise in answers.

This comparison can help operators to improve the process of collecting feedback answers. A design improvement could be to perform survey in a non neutral environment. For example, it could be conducted online at the precise date when a network issue is suspected for a specific user. About predictive improvements, gathering additional data should help to refine prediction model, especially for complaints data.

## REFERENCES

Alan, D. S., & Basu, K. (1994). Customer loyalty: Toward an integrated conceptual framework. *Journal of the Academy of Marketing Science*, 22(2), 99–113. doi:10.1177/0092070394222001

Bozdogan, H. (1987). Model selection and Akaikes information criterion (AIC): The general theory and its analytical extensions. *Psychometrika*, 52(3), 345–370. doi:10.1007/BF02294361

Cortes, C., & Mohri, M. (2004). AUC optimization vs. error rate minimization. *Advances in Neural Information Processing Systems*, 16, 313–320.

Dietterich, T. G. (2000). Ensemble methods in machine learning. In J. Kittler & F. Roli (Eds.), *Multiple classifier systems* (pp. 1–15). Berlin: Springer. doi:10.1007/3-540-45014-9_1

Fowler, F. J. J. (2013). *Survey research methods*. Thousand Oaks, CA: Sage publications.

Goodman, J., & Newman, S. (2003). Understand customer behavior and complaints. *Quality Progress, 36*(1), 51–55.

Guerrero, V. M., & Johnson, R. A. (1982). Use of the Box-Cox transformation with binary response models. *Biometrika, 69*(2), 309–314. doi:10.1093/biomet/69.2.309

Hadden, J., Tiwari, A., Roy, R., & Ruta, D. (2006). Churn prediction using complaints data. *Proceedings of World Academy of Science, Engineering and Technology*.

Hand, D. J. (2009). Measuring classifier performance: A coherent alternative to the area under the ROC curve. *Machine Learning, 77*(1), 103–123. doi:10.1007/s10994-009-5119-5

Hand, D. J., & Anagnostopoulos, C. (2013). When is the area under the receiver operating characteristic curve an appropriate measure of classifier performance? *Pattern Recognition Letters, 34*(5), 492–495. doi:10.1016/j.patrec.2012.12.004

Hosmer, J. D. W., & Lemeshow, S. (2004). *Applied logistic regression*. New York: John Wiley & Sons.

Huang, J., & Ling, C. X. (2005). Using AUC and accuracy in evaluating learning algorithms. *Knowledge and Data Engineering. IEEE Transactions, 17*(3), 299–310.

Hurvich, C. M., & Tsai, C. (1989). Regression and time series model selection in small samples. *Biometrika, 76*(2), 297–307. doi:10.1093/biomet/76.2.297

Kuhn, M. (2008). Building predictive models in R using the caret package. *Journal of Statistical Software, 28*(5), 1–26. doi:10.18637/jss.v028.i05 PMID:27774042

Liaw, A., & Wiener, M. (2002). Classification and Regression by randomForest. *R News, 2*(3), 18-22.

Lobo, J. M., Jiménez-Valverde, A., & Real, R. (2008). AUC: A misleading measure of the performance of predictive distribution models. *Global Ecology and Biogeography, 17*(2), 145–151. doi:10.1111/j.1466-8238.2007.00358.x

Oliver, R. L. (1997). *Satisfaction: A behavioral perspective on the customer*. New York: Irwin McGraw Hill.

Oliver, R. L. (1999). Whence consumer loyalty? *Journal of Marketing*, *63*, 33–45. doi:10.2307/1252099

Qian, H. (2011). China Mobile satisfaction survey analysis. *Jiangsu Science and Technology Information*, *9*, 27–29.

Reichheld, F. F. (2003). The one number you need to grow. *Harvard Business Review*, *81*(12), 46–55. PMID:14712543

Rossi, P. H., Wright, J. D., & Anderson, A. B. (Eds.). (2013). *Handbook of survey research*. Academic Press.

Saeys, Y., Inza, I., & Larrañaga, P. (2007). A review of feature selection techniques in bioinformatics. *Bioinformatics (Oxford, England)*, *23*(19), 2507–2517. doi:10.1093/bioinformatics/btm344 PMID:17720704

Sakia, R. M. (1992). The Box-Cox transformation technique: A review. *The Statistician*, *41*(2), 169–178. doi:10.2307/2348250

Sing, T., Sander, O., Beerenwinkel, N., & Lengauer, T. (2005). ROCR: Visualizing classifier performance in R. *Bioinformatics (Oxford, England)*, *21*(20), 3940–3941. doi:10.1093/bioinformatics/bti623 PMID:16096348

Suykens, J. A. K., & Vandewalle, J. (1999). Least squares support vector machine classifiers. *Neural Processing Letters*, *9*(3), 293–300. doi:10.1023/A:1018628609742

Yu, M. (2014). *China Mobile, customer satisfaction and promotion strategies for network companies* (Unpublished dissertation). Chongqing University, China.

# Chapter 5
# Predicting 4G Adoption with Apache Spark:
## A Field Experiment

**Mantian (Mandy) Hu**
*The Chinese University of Hong Kong, China*

## ABSTRACT

*Companies have long realized the value of targeting the right customer with the right product. However, this request has never been so inevitable as in the era of big data. Thanks to the tractability of the customers' behavior, the preference information for each individual is collected and updated by the firm in a timely fashion. In this study, we developed a targeting strategy for telecommunication companies to facilitate the adoption of 4G technology. Utilizing the most up to date machine learning technique and the information about individual's local network, we set up a prediction model of consumer adoption behavior. We then applied the model to the real world and conduct field experiment. We worked with the largest telecommunication company in China and used Apache Spark to analyze the data from the complete customer based of a 2nd tie city in eastern China. In the experiment group, we asked the company to use the list we generated as the targets and in the control group, the company used the existing targeting strategy. The results demonstrated the effectiveness of the proposed approach comparing to existing models.*

DOI: 10.4018/978-1-5225-1750-4.ch005

## INTRODUCTION

With the development of TD-LTE mobile network, more 4G base stations are launched and more mobile users change to use 4G (the fourth generation mobile communication and technology service) network for its better coverage and high speed. How to guide 2/3G users to use 4G network is a major challenge for mobile telecom operators. The market department usually targets every user to do the transformation, but this method has a lack of accuracy, efficiency and is quite expensive to accomplish.

This article points an efficient method to improve the 4G transfer rate by considering peer influence from local network data. User's choice is influenced by people, especially by their close friends. This phenomenon is called peer influence. Here local network refers to a network produced by interactions and connections for each individual. From sociology side, every mobile user has its own communication circle. While from mathematical perspective, local network is an ordinary network and every mobile user can be taken as a node in the network. According to calls and messages history, the connections between user to other users are created. Therefore, we can analyze individual behaviors from the network.

Our results show that using machine learning algorithms can increase the 4G transfer rate, and local network data can be used to improve the performance.

The remainder of this paper is organized as follows: the section titled Context Data shows the existing research related to this paper; the section titled experimental design describes the context of the experiment and the dataset; the section titled modeling progress and algorithm details the design of the experiment and related machine learning methods; the results section shows the main results of this paper which we conclude in the last section.

## RESEARCH BACKGROUND

The existence of 4G networks in today's technology-driven society is an important label of advancement and change. 4G networks are designed to improve wireless capabilities, network speeds and so on. New technology always brings opportunities and challenges (Gobjuka, 2009). One of the challenge is to attract more users to use 4G network. More 4G users means more important status in telecommunication market for telecom operators. As the competitions among operators are increasing, a creative and efficient market strategy is needed to attract more 4G users.

There are already some researches to learn behaviors of 4G users. A structural equation modeling (SEM) is conducted (Subramanian, 1994) and the results show that attitude towards the use of 4G network and the perceived usefulness of it are significant facilitators of users intention to adopt this technology. A study (Rawashdeh, 2015) aimed to examine the relationship between the perceived usefulness, perceived ease of use, perceived entertainment, attitude and the users' intention toward using the 4G wireless mobile services. The findings reveal that the combination of them are together responsible for determining the users' intention to use 4G network, factor analysis, correlation and regression analysis are used in the paper.

User choice is influenced by actions taken by others (Leenders, 2002). In particular by actions of close friends (Lee, Hosanagar, & Tan, 2015). There are a lot of underlying research explaining why this is the case, such as information transmission (Katz & Lazarsfeld, 1955), competition (Burt, 1987) and conformity (Menzel, 1960). These are usually called peer influence in the literature [5]. Researchers have completed that peer influence plays a important role in many different applied fields such as in the diffusion of new drugs (Coleman, Katz, & Menzel, 1966), messaging services (Aral, Muchnik, & Sundararajan, 2009), and in the sales of online books (Chevalier & Mayzlin, 2006). Peer influence has also been shown to shape knowledge diffusion and academic performance (Sacerdote, 2001), opinions and product reviews (Muchnik, Aral, & Taylor, 2013).

This paper applies peer influence research and uses outcomes of a random experiment to estimate the 4G transfer rate by a 4G adoption prediction model – using the most up to date machine learning algorithms. Furthermore, some feature selection methods are used in the experiments, which turns out to be great indicators from business side. Finally, the framework of this analysis is replicable in other companies that attempt to measure its impact at user level.

## CONTEXT AND DATA

Our data come from Wuxi Mobile Communications Corporation (Jiangsu province, China). Below, we describe the current market conditions at Wuxi and the data available to us.

### 1. Market Condition

Attracting more 4G users is carried out as part of Wuxi Mobile's Customer Relationship Management (CRM) operations. There are two steps listed below:

- **User Satisfaction Management:** Wuxi Mobile offers different services with low marginal cost to different consumption users, which may potentially increase product utilization and customer satisfaction. Additionally, Wuxi

Mobile often provides free 4G flux activities, purchases 4G flux with half price and other promotions.

- **New User Marketing:** for new users, Wuxi Mobile sends a lot of free 4G flux. For those 2/3G users, it provides 4G services without changing user's phone number.

4G product promotion has been lasted over 1 years at Wuxi mobile. Until now, half of their users already transferred to use 4G network, while the 4G transfer rate of existing users is only 4.8%. The challenge is imminent and Wuxi Mobile needs to find why some users transfer while others don't transfer. The goal is to identify likely transfer users in advance so client could have enough time to recommend proper 4G products. As a result of this collaboration, we developed a 4G adoption prediction model for Wuxi Mobile and a process to measure its efficiency.

## 2. Overview of Datasets

Wuxi Mobile provides us two types of data: business analysis data and local network data. These information have been used to build a 4G adoption prediction model.

Business analysis data contain a snapshot of all products subscribed for each user and their monthly consumption. The demographic information and monthly billing information can also get from this kind of data. Finally, we also have tenure information from the dates when users join Wuxi Mobile and use their phone terminal.

Local network data include users' calling history and message history. Those data are used to create local network. According to these data, a telecom communication network can be generated, and from this network, we can analyze each user's behaviors.

All the datasets are collected by Wuxi Mobile between July 2015 to September 2015. The business analysis data are randomly selected from all users from Wuxi Mobile, which has 1 million unique records with 4G users and non-4G users. The local network data contain over 300 million calls and text messages, their status whether using 4G network or not is also included in those data. We use this information to define an undirected network of communications across users. In order to build this network, we match all phone numbers to their contacts and delete all records with no match in business analysis data. The edge between two users is included in this network if all of the following conditions are met:

1.    There are call in and call back history between them;
2.    At least one of the calls takes place during the weekend;
3.    There are at least three calls or messages between them in two months.

Thus, the graph is close to user real connections. The size of this data is about 3GB. Consequently, we need a big data processing platform to run this algorithm.

The Apache Spark (Spark Prgramming Guide, 2015) is a fast and general-purpose cluster computing system. It provides high-level APIs in Java, Python and R, and an optimized engine that supports general execution network.

## EXPERIMENTAL DESIGN

Our research has two interests. One interest is we wanted to know the performance of Wuxi Mobile's own management strategy. For the other interest, we wanted to know whether using local network data and machine learning method could increase the performance of 4G transfer rate. Because there are many other factors which are likely correlated with the network created above, the best way is to do a random experiment to ensure identification. Below is the experiment we designed.

From local network data, we need to know which user has more 4G user connector, how much time duration they speak to these 4G user and other information related to these connectors. We assume more connection to 4G users will lead to a higher transfer probability. Then we apply a machine learning method to get the probability to use 4G network. With these double standards, we can get our targeting list and take this group as test group. In the control group, Wuxi Mobile used their existing targeting strategy to do the promotion.

## MODELING PROCESS AND ALGORITHM

In this section, the process to do this model is described. Our aim is to distinguish 4G users from non 4G users. Since this problem is a binary problem, classification algorithms are used. The first step is to prepare data: in this step, training data, validation data and test data are separated from original data. The second step is variable dealing: in this step, variables are divided into two types, one is categorical variables and the other is continuous variables. The third step is to choose the predictive features, which is a crucial step in the model and different statistic method are used. Finally, different machine learning tools tested for classification are introduced.

### 1. Data Preparing

It has been mentioned that there are two types of data used in this paper. One is business analysis data and the other is local network data. Model data prepare process will be introduced in this part.

*Table 1. Sample data distribution*

| Data Name | Distribution | |
|---|---|---|
| | Records | 4G User per |
| Training Data | 48,908 | 46.41% |
| Validation Data | 21,092 | 46.34% |
| Test Data | 16,000 | 46.36% |

After we build network according to the restrictions described in section III, the resulting local network contains 0.15 million nodes and 62,000 edges. Each node has following information, such as the number of connected 4G users, call duration, time with 4G users and so on. The model data are ready after combining this data with business analysis data.

The model data contains 46% 4G users, then we randomly select 70,000 records from the model data to build model and 16,000 records to test the model result. Table 1 shows sample data distribution.

## 2. Feature Extraction and Transformation

We need to do feature extraction and transformation after finishing data splitting. In this paper, the features are divided into two types: categorical and continuous variables.

For categorical variables, we need to do dummy variable processing. Choose the most frequent 2 or 4 values to keep, while missing value and other values are ignored.

For continuous variables, there are two important steps: one is missing value dealing and the other is outlier dealing. We can choose value between 1%-99% or 5%-95% for a variable. Then plot the curve with dependent variable, fill the missing value to make the curve monotone decreasing or increasing. If the curve is irregular, we think this feature is not significant in the model. Transform the feature with log, square and square-root format.

Since there are at least 100-500 variables need to do the extraction and transformation, we have made this process automatically.

## 3. Feature Selection

Feature selection methods are introduced in this section. There are three methods introduced:

- **Forward Selection:** Features are selected from one variable into the model. Calculate the F statistic and P value according to the squares of deviations (SS2) of features in the model. Feature can be selected if the P value is less than the threshold. The limitation of the forward selection method is if the threshold is set as a small value, there may be no features in selected list. If too big, selected features at beginning can't be removed later.

- **Backward Selection:** Starting with all the features included in training data. The F statistic and P value for each feature will be calculated. Feature is selected when P value is smaller than the threshold. Otherwise, delete features from the one with highest P value until there is no feature to remove. The limitation for this method is similar with forward selection, the threshold shouldn't set too big or too small, otherwise there will be no variables no exclude and can't be selected in the model.

- **Stepwise Selection:** This method is combined with forward selection and backward selection. The features will be selected like forward selection methods which is decided by the F statistics and threshold of P value. When features are selected, it will delete features as backward selection method. This method is better than forward selection and backward selection method, but has its own limitations: first, when there are M features in the selection list, this combination is not the best choice to the M+1 feature. Second limitation is it only take F statistics to select or delete features as a standard.

After these three steps, the initial feature list is set the union of those selected features. Then, we use logistic regression to see each feature's Variance Inflexion Factor (VIF) value (no more than 2) and P value (no more than 0.05) to do further selection.

## 4. Machine Learning Algorithm for Classification Model

Use 4G network or not is a typical classification topic. There are many classification algorithms and the main objective is to find a classify machine or function to separate 4G network users from non 4G users. There are three classification method used in this paper.

- **Logistic Regression ( Hosmer & Lemeshow, 2004 ):** Logistic regression us a machine learning algorithm which is commonly used in the industry, and it is used to estimate the probability of a certain thing. The algorithm is fast and able to deal with high dimension data. However, it can't catch complex non-linear situation.

- **Random Forest (RF) ( Liaw & Wiener, 2002 ):** Random Forest algorithm is an efficient classification algorithm. After training step, it can directly show the importance of each features according to the information gain and detect the interaction between features. In addition, based on its white box characteristic, if given a 4G user, it's easy to launch the corresponding logical expressions and explain from the generated decision trees.

- **Support Vector Machine (SVM) ( Suykens & Vandewalle, 1999 ):** SUPPORT vector machine has many advantages in solving small sample, nonlinear and high dimensional pattern recognition. A support vector is constructed by a hyper plane, in high or infinite dimensional space, which can be used for classification, regression, or other tasks in a set of hyper planes. The kernel trick is an important concept which perform non-linear classifications.

All above methods are used to do this model. After the features are selected, scoring the training data and validation data to see if there exists oversample situation with the equation we get. Let L represents the list of users sorted by decreasing order of 4G transfer probability, and divide L into ten groups. If the performance on each group is decreasing and consistent on two data, then it will be a good model. Finally, we compare the AUC (Huang & Ling, 2005) and AIC (Bozdogan, 1987; Burnham & Anderson, 2004) to select a best one.

## RESULTS

In this part, we will mainly show our results of this model:

1. List two to three selected features;
2. Show the performance on training and validation part;
3. Show the performance on model data and test data;
4. Compare results between experiments.

## 1. Selected Features

The average of last three month's charging flux (avg_sfll) is one of the most important features, because of its positive correlation with dependent variable. It can be inferred that in past three months, people who use more charge flux may have a higher probability to use 4G network.

User terminal usage time is another important feature. The longer time usage, the less probability to use 4G network. In other words, due to the universal 4G mobile market, people who changed their mobile are more easily to use 4G network.

It's worth to be mentioned that the local network we build is very efficient. User who has more 4G connected contactors has a high transfer rate.

It's noted that the importance of each feature is calculated based on standard estimate.

## 2. Performance on Training and Validation Data

Sorting the training and validation data by decreasing order of 4G transfer possibility, divided the data into ten groups, then we will get a decreasing 4G transfer rate for each group if the model is good enough. From Tables 2 and 3, we can say this model is a good model and there is no over-fitting problems since the performance on each group is quite consistent on training and validation data.

## 3. Performance on Model Data and Test Data

Apply the same method on the model data and test data, we can get a quite good result. But here we focus on the comparison between predicted value and real value. On model data, we compare real and predicted 4G transfer rate while on test data, predicted and real 4G number is compared:

The overall model performance is quite visible. The trend between real and predicted 4G transfer rate is consistent on model data and decreasing on each group. What's the model performance on data that outside of the model data? In Table 4

*Table 2. Model performance on training data*

| Training Group | Model Performance on Training Data | | |
|:---:|:---:|:---:|:---:|
| | Universe | # of 4G Users | % of 4G Users |
| 1 | 4,891 | 4,846 | 99.08% |
| 2 | 4,891 | 4,723 | 96.57% |
| 3 | 4,890 | 4,299 | 87.91% |
| 4 | 4,891 | 3,470 | 70.95% |
| 5 | 4,891 | 2,562 | 52.38% |
| 6 | 4,891 | 1,446 | 29.56% |
| 7 | 4,891 | 686 | 14.03% |
| 8 | 4,890 | 311 | 6.36% |
| 9 | 4,794 | 231 | 4.82% |
| 10 | 4,988 | 126 | 2.53% |

*Table 3. Model performance on valadation data*

| Validation Group | Model Performance on Validation Data | | |
|---|---|---|---|
| | Universe | # of 4G Users | % of 4G Users |
| 1 | 2,109 | 2.093 | 99.24% |
| 2 | 2,109 | 2,036 | 96.54% |
| 3 | 2,110 | 1,832 | 86.82% |
| 4 | 2,109 | 1,505 | 71.36% |
| 5 | 2,109 | 1,092 | 51.78% |
| 6 | 2,109 | 646 | 30.63% |
| 7 | 2,109 | 304 | 14.41% |
| 8 | 2,110 | 121 | 5.73% |
| 9 | 2,100 | 98 | 4.67% |
| 10 | 2,118 | 49 | 2.31% |

*Figure 1. Real vs. predicted value on model data*

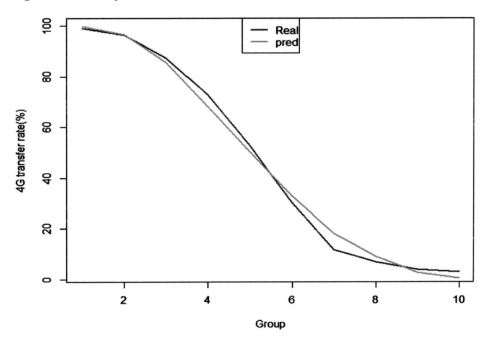

*Table 4. Model performance on test data*

| Test Data Group | Mode Performance on Test Data | | |
|---|---|---|---|
| | Universe | Real 4G Users | Predicted 4G Users |
| 1 | 1,556 | 1,548 | 1,551 |
| 2 | 1,647 | 1,595 | 1,584 |
| 3 | 1,560 | 1,396 | 1,325 |
| 4 | 1,483 | 1,122 | 1,000 |
| 5 | 1,778 | 1,015 | 909 |
| 6 | 1,654 | 523 | 561 |
| 7 | 1,707 | 259 | 328 |
| 8 | 1,734 | 118 | 146 |
| 9 | 1,701 | 64 | 45 |
| 10 | 1,790 | 61 | 6 |

*Figure 2. Real vs. predicted value on model data*

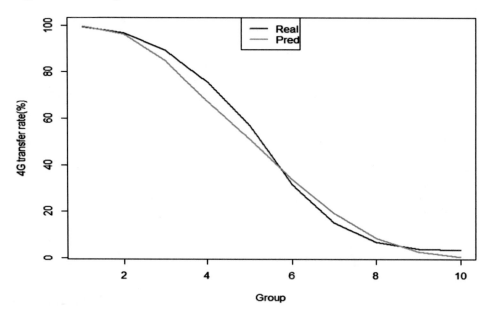

we show the model performance on test data and use the minimum value as cutoff point to divide the group of test data.

The comparison between real and predicted 4G transfer rate is like performance on model data is shown in Figure 2.

*Table 5. Data distribution*

| Dimension | Data Distribution |
|---|---|
| | Universe |
| 4G Terminal | 14,291 |
| 2/3G Terminal | 337,892 |

*Table 6. 4G terminals*

| Group | 4G Adopter | |
|---|---|---|
| | Number | Percentage |
| Control | 443 | 30.49% |
| Test | 290 | 20.3% |

*Table 7. 2/3G terminals*

| Group | 4G Adopter | |
|---|---|---|
| | Number | Percentage |
| Control | 4,781 | 10.65% |
| Test | 2,800 | 8.2% |

From Figures 1 and 2, the number of real transfer rate is very close to the number of prediction. Through calculation, the accurate rate for the model is as high as 93.2%.

## 4. Experiment Results Comparison

Wuxi Mobile sends us 350,000 users who didn't use 4G network in October 2015, we scored these persons and got each user's probability to use 4G network.

Wuxi Mobile see users' 4G transfer rate on two dimensions: one type is users who use 4G terminal but don't use 4G network, the other is users who still use 2/3G terminal. We also divided the target persons on these two dimensions, and Table 5 shows distribution of these two dimensions.

As the design of the experiment, a control and test group are compared, and both of these two groups choose the top 10% users with high transfer probability.

One month later, the results come up as shown in Tables 6 and 7.

From Tables 6 and 7, it's quite obvious that the control group beat the test group on the final 4G adopter.

## CONCLUSION AND IMPROVEMENTS

Our marketing strategy is based on users communication circle, taking advantage of the peer influence and machine learning algorithms to improve the marketing efficiency. This experiment is conducted on two datasets, one is business analysis data while the other is local network data. For the massive datasets, the Apache Spark was chosen for its good scalability, effectiveness and efficiency. The result shows that our methodology has a great increase on 4G adopter compared with Wuxi Mobile's own targeting strategy.

New models could help to improve targeting results. For example, adding differentiation on the dimension of users, using 4G terminals but don't use 4G network, the other is users who still use 2/3G terminal. For users who don't use 4G network, we can also do a clustering model to differentiate the users. Making a profiling on each cluster and summarizing the reason why users don't use 4G network, which can help Wuxi Mobile do different targeting policy to different cluster.

## REFERENCES

Aral, S., Muchnik, L., & Sundararajan, A. (2009). Distinguishing influence based contagion from homophily-driven diffusion in dynamic networks. *Proceedings of the National Academy of Sciences of the United States of America, 106*(51), 21544–21549. doi:10.1073/pnas.0908800106 PMID:20007780

Bozdogan, H. (1987). Model selection and Akaikes information criterion(AIC): The general theory and its analytical extensions. *Psychometrika, 52*(3), 345–370. doi:10.1007/BF02294361

Burnham, K. P., & Anderson, D. R. (2004). Multimodel inference understanding AIC and BIC in model selection. *Sociological Methods & Research, 33*(2), 261–304. doi:10.1177/0049124104268644

Burt, R. S. (1987). Social contagion and innovation: Cohesion versus structural equivalence. *American Journal of Sociology, 92*(6), 1287–1335. doi:10.1086/228667

Chevalier, J. A., & Mayzlin, D. (2006). The effect of word of mouth on sales: Online book reviews. *JMR, Journal of Marketing Research, 43*(3), 345–354. doi:10.1509/jmkr.43.3.345

Coleman, J. S., Katz, E., & Menzel, H. (1966). *Medical innovation: A diffusion study*. Bobbs-Merrill Co.

Gobjuka, H. (2009). *4G wireless networks: Opportunities and challenges*. arXiv preprint arXiv: 0907.2929

Hosmer, D. W. Jr, & Lemeshow, S. (2004). *Applied logistic regression.* John Wiley & Sons.

Huang, J., & Ling, C. X. (2005). Using AUC and accuracy in evaluating learning algorithm. *IEEE Transactions on Knowledge and Data Engineering, 17*(3), 299–310. doi:10.1109/TKDE.2005.50

Katz, E., & Lazarsfeld, P. F. (1955). *Personal Influence, The part played by people in the flow of mass communications.* Transcation Publishers.

Lee, Y. J., Hosanagar, K., & Tan, Y. (2015). Do I follow my friends or the crowd? Information cascades in online movie ratings. *Management Science, 61*(9), 2241–2258. doi:10.1287/mnsc.2014.2082

Leenders, A. J. (2002). Modeling social influence through network autocorrelation: Constructing the weight matrix. *Social Networks, 24*(1), 21–47. doi:10.1016/S0378-8733(01)00049-1

Liaw, A., & Wiener, M. (2002). Classification and regression by RandomForest. *R News, 2*(3), 18-22.

Menzel, H. (1960). Innovation, integration, and marginality: A survey of physicians. *American Sociological Review, 25*(5), 704–713. doi:10.2307/2090143

Muchnik, L., Aral, S., & Taylor, S. J. (2013). Social influence bias: A randomized experiment. *Science, 341*(6146), 647–651. doi:10.1126/science.1240466 PMID:23929980

Rawashdeh, A. (2015). Adoption of 4G mobile services from the female student's perspective: Case of Princess Nora University. *Malaysian Online Journal of Educational Technology, 3*(1), 12–27.

Sacerdote, B. (2001). Peer effects with random assignment: Results for dartmouth roommates. *The Quarterly Journal of Economics, 116*(2), 681–704. doi:10.1162/00335530151144131

Spark Programming Guide – Spark 1.6.0 Documentation. (2015). Retrieved from https://spark.apache.org/docs/1.6.0/programming-guide.html

Subramanian, G. H. (1994). A Replication of Perceived Usefulness and Perceived Usefulness and Perceived Ease of Use Measurement. *Decision Sciences, 25*(5-6), 863–874. doi:10.1111/j.1540-5915.1994.tb01873.x

Suykens, J. A. K., & Vandewalle, J. (1999). Least squares support vector machine classifiers. *Neural Processing Letters, 9*(3), 293–300. doi:10.1023/A:1018628609742

# Chapter 6
# Mining of Leaders in Mobile Telecom Social Networks

**Mantian (Mandy) Hu**
*The Chinese University of Hong Kong, China*

## ABSTRACT

*In the age of Big Data, the social network data collected by telecom operators are growing exponentially. How to exploit these data and mine value from them is an important issue. In this article, an accurate marketing strategy based on social network is proposed. The strategy intends to help telecom operators to improve their marketing efficiency. This method is based on mutual peers' influence in social network, by identifying the influential users (leaders). These users can promote the information diffusion prominently. A precise marketing is realized by taking advantage of the user's influence. Data were collected from China Mobile and analyzed. For the massive datasets, the Apache Spark was chosen for its good scalability, effectiveness and efficiency. The result shows a great increase of the telecom traffic, compared with the result without leader identification.*

## INTRODUCTION

Number of smart phones users increases quickly with the growth of 4G LTE mobile networks. How to improve traffic consumption of mobile users is an important task for telecom operators. In recent years, competition between them has become much

DOI: 10.4018/978-1-5225-1750-4.ch006

more intense. Traditionally, they used to make advertising campaigns to attract users. However, this method has a lack of accuracy, efficiency, and is expensive to accomplish.

Social network refers to a relationship system induced by interaction between individuals. It focuses on interactions and connections between the different users. From social network's perspective, every mobile user has its own communication circle. According to messages and calls of users, a telecom communication connections can be generated. Then, individual behaviors can be analyzed in the communication network to identify leaders. These leaders influence opinions and decisions of others users. They have been used in marketing, public opinion analysis, disease control and many other fields. For example, Lam and Wu (2009) find a subset of influential users in the social network of eBay and use it to rank potential buyers for viral marketing.

## SOCIAL NETWORK BEHAVIOR

In sociology, an individual's behavior is not only determined by his character, but also by his friends' behavior. Then, when we analyze an individual behavior, we need to consider his friends effect. Godinho de Matos, Ferreira, and Krackhardt (2014) estimate peer influence in the diffusion of iPhone 3G across different communities. They find that if one's friend has iPhone 3G, there is a greater probability for him to switch to this cell phone when changing it. Pushpa (2012) construct a telecom social network, and find that the efficiency to predict customer churn based on social network has a good increase.

In communication networks, users build relationships each other by mobile communication like calls and messages. This relationship between individuals can been seen as a graph, each node of a graph representing a user. An example of such graph is depicted on Figure 1.

Often, if the original network is large, it is necessary to construct or identify subpopulations from this large network, before using them for analysis. In complex networks, there are many ways to detect a community. Newman (2004) proposes a modularity spectral optimization, uses modularity as an evaluation criteria and detects subpopulations by maximization modularity. Clauset, Newman, and Moore (2005) design a greedy maximization of the local modularity algorithm. This algorithm is considered to be one of the fastest community extraction algorithms. For the communication circle's character, we use the snowball sampling method to extract the subgraph from the whole mobile users.

In statistics research, snowball sampling (Goodman, 1961) is a non-probabilistic sampling technique and is a useful tool for building networks and increasing the

*Figure 1. Communication network example*

number of participants. We use a snowball sampling method to get the seed user's contacts, and his contacts' contacts. Therefore, we obtain the seed user's two-layers contacts using this sampling method. An example of such two-layers contacts is shown on Figure 2.

After having performed community detection, we use the Internal-External Ratio index (Zhang, 2011) to optimize this community, eliminating the low coher-

*Figure 2. Two-layers snowball sampling of a user*

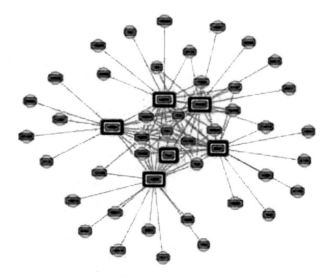

ent users. Consequently, users in this circle are closer to each other and relationship between different circle becomes sparse.

## IDENTIFICATION OF INFLUENTIAL INDIVIDUALS

In complex networks (Havlin & Cohen, 2010; Newman, 2003), the degree of a node in a network is the number of connections it has to other nodes. Usually, network is directed, which means that edges are directed from one node to another node. In this case, a node has two different degrees: the in-degree, which refers to the number of nodes linked to this node; and the out-degree, which is the number of outgoing edges from this node. If the number of nodes linked to this node is high, it means this node has great influence on other nodes in this network.

When we analyze complex networks, various parameters must be taken into account such as: clustering coefficient, degree centrality, closeness centrality, eigenvector centrality, etc. Those features can be extracted and easily exploited to determine the most influential node in the community. Liebig and Rao (2014) take structured clusters into account, and use bipartite clustering coefficient to find important nodes across communities. Opsahl, Afneessens, and Skvoretz (2010) measure node centrality by combining degree, closeness, and betweenness and use node centrality to calculate the influential nodes.

The PageRank is an algorithm created by Sergey Brin and Larry Page (1998). It was used by Google Search to rank websites in their search engine results. It is a way of measuring the importance of website pages.

*Figure 3. Node's relationship example*

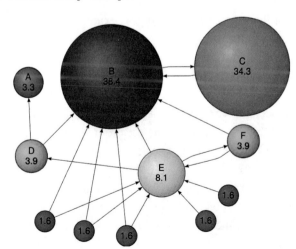

*Figure 4. Link relationship example*

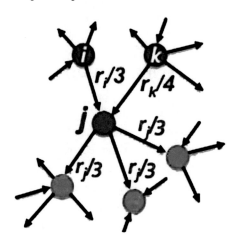

In this algorithm, one node's importance is determined by its neighbour's importance as well as numbers linked to this node. On Figure 3, we can see that the more linked by other node, the more important the node is. Besides, if its neighbours' importance is high, the node is more important.

On Figure 4, we use symbol r to represent the node's importance, and get the importance of node j by computing:

$$r_j = r_i/3 + r_k/4. \tag{1}$$

According to the link relationship in the social network, we can get every node's expression like node j. From Figure 5, we get the adjacency matrix called A.

Then, we use the equation Y = A*Y' to update node importance vector Y. As the node number of this community is 3 in this example, we can give to every node the initial importance value 1/3.

After several times iterations, the importance value of each node converge to 6/15, 6/15, 3/15, as shown on Figure 6. We can see that nodes y and a have more weight than node m.

In the telecom social network, we think of users as nodes, and use communication and message records as links between the different users. The R Package "igraph" (Package igraph, 2014) is used for network analysis and visualization.

We use the PageRank algorithm to compute every node's importance. After 100 iterations, we get the importance of all users, and leaders are identified (in the example shown on Figure 7, user labeled with number 3 is identified).

Users' data have been collected from China Wuxi Mobile. We select those whose 3G flux per month are between 50MB and 100MB as root users. The number of

*Figure 5. Adjacency matrix*

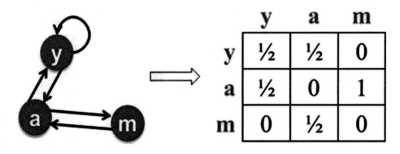

|   | y | a | m |
|---|---|---|---|
| y | ½ | ½ | 0 |
| a | ½ | 0 | 1 |
| m | 0 | ½ | 0 |

*Figure 6. Node's importance*

$$\begin{pmatrix} r_y \\ r_a \\ r_m \end{pmatrix} = \quad \begin{matrix} 1/3 & 1/3 & 5/12 & 9/24 \\ 1/3 & 3/6 & 1/3 & 11/24 \\ 1/3 & 1/6 & 3/12 & 1/6 \end{matrix} \quad \cdots \quad \begin{pmatrix} 6/15 \\ 6/15 \\ 3/15 \end{pmatrix}$$

Iteration 0, 1, 2, ...

these root users is about 220,000 and data size of the users' two-layer network is 3.6GB, for about 4 million users. Thus, we need a big data processing platform to run this algorithm.

The Apache Spark (Spark Prgramming Guide, 2015) is a fast and general-purpose cluster computing system. It provides high-level APIs in Java, Scala, Python and R, and an optimized engine that supports general execution graphs. The GraphX is a component in Spark for graph and graph-parallel computations. It models a "property graph", which is an multi directed, attributed graph and includes a growing collection of graph algorithms and builders to make graph computations.

If the numbers in one circle is too small, there is no need for us to market to those customers. Thus, in the pruning phase, circles with a node number smaller than 50 are deleted. Then, we use the GraphX platform to compute the leaders in these communication circles. Finally, we get about 60,000 leaders whose average data-allowance is above 500M. They are all our target marketing users and we exploit their influence on their friends to promote our business.

The China Wuxi Mobile has a community named Commune. Members of this Commune have a preferential flow of consumption. In the marketing phase, China

*Figure 7. Call records*

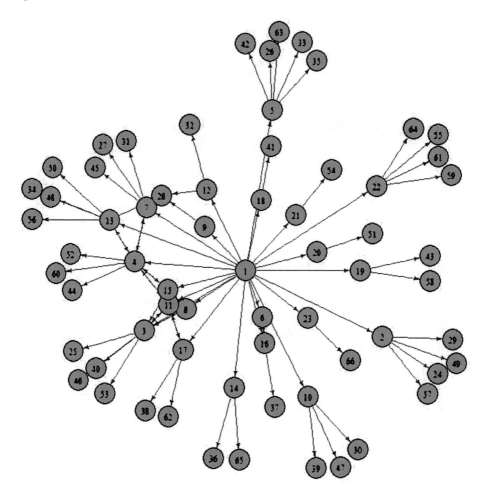

Wuxi Mobile gives 100 Commune Credits to all the 60,000 leaders and let them as the Registered users of the Commune. Those leaders are allowed to share credits with their friends. In addition, if his friend register in the Commune and his mobile traffic increase in the next month, we give to the leaders and his corresponding friends another 100 Commune Credits. The result can be seen on Figure 8.

In Figure 8, "Mobile traffic" represents the mobile flow consumption of users in Commune. "Registered users" represents the number of users registered in Commune. "Active users" represents the number of users using APPs to generate data. We observe that from June to September, when marketing strategy haven't been developed yet, the growth rate of mobile data increased slowly, as for both "Registered users" and "Active users". Then, by giving credits to the leaders in

*Figure 8. Users and traffic variation*

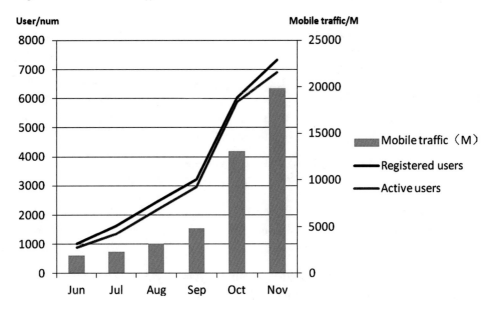

September, we observe that these three items have a rapid increase in October and November. By comparison with September, in October, the growth rate of mobile traffic has raised from 36% to 150%, the registered users from 50% to 87.8% and the active users from 36% to 150%, leading to a rapid growth. In November, the growth becomes slower compared with October.

## CONCLUSION

Our marketing strategy is based on communication circle. It takes advantage of the leader' influence on his peers to improve the marketing efficiency. In one's communication circle, the user behavior data not only help us to depict one's character, but also his communication circle's character. Therefore, we can implement different marketing project for different groups. If one's group has a high-level, we can recommend high-end mobile phone to them. Compared with ordinary marketing strategy, the method based on communication network and leader's influence is more precise and effective.

# REFERENCES

Brin, S., & Page, L. (1998). The anatomy of a large-scale hypertextual Web search engine. *Computer Networks and ISDN Systems*, *30*(1-7), 107–117. doi:10.1016/S0169-7552(98)00110-X

Clauset, A., Newman, M. E. J., & Moore, C. (2005). Finding community structure in very large networks. *Physical Review E: Statistical, Nonlinear, and Soft Matter Physics*, *70*(6), 066111. doi:10.1103/PhysRevE.70.066111 PMID:15697438

Cohen, R., & Havlin, S. (2010). *Complex Networks: Structure, Robustness and Function*. Cambridge University Press. doi:10.1017/CBO9780511780356

Godinho de Matos, M., Ferreira, P., & Krackhardt, D. (2014). (Forthcoming). Peer Influence in the Diffusion of the iPhone 3G over a Large Social Network. *Management Information Systems Quarterly*.

Goodman, L. A. (1961). Snowball sampling. *Annals of Mathematical Statistics*, *32*(1), 148–170. doi:10.1214/aoms/1177705148

Lam, H. W., & Wu, C. (2009). Finding influential ebay buyers for viral marketing a conceptual model of BuyerRank. In *2009 International Conference on Advanced Information Networking and Applications* (pp. 778-785). IEEE. doi:10.1109/AINA.2009.36

Liebig, J., & Rao, A. (2014). Identifying Influential Nodes in Bipartite Networks Using the Clustering Coefficient. In *2014 Tenth International Conference on Signal-Image Technology and Internet-Based Systems (SITIS)* (pp. 323-330). IEEE. doi:10.1109/SITIS.2014.15

Newman, M. E. J. (2003). The structure and function of complex networks. *SIAM Review*, *45*(2), 167–256. doi:10.1137/S003614450342480

Newman, M. E. J. (2004). Detecting community structure in networks. *The European Physical Journal B-Condensed Matter and Complex Systems*, *38*(2), 321–330. doi:10.1140/epjb/e2004-00124-y PMID:15244693

Opsahl, T., Agneessens, F., & Skvoretz, J. (2010). Node centrality in weighted networks: Generalizing degree and shortest paths. *Social Networks*, *32*(3), 245–251. doi:10.1016/j.socnet.2010.03.006

Package igraph, Version 0.7.1. (2014). Retrieved from http://igraph.org/2014/04/21/igraph-0.7.1-c.html

Pushpa, S. (2012). G.: An Efficient Method of Building the Telecom Social Network for Churn Prediction. *International Journal of Data Mining & Knowled Management Process*, 2(3), 31–39. doi:10.5121/ijdkp.2012.2304

Spark Programming Guide – Spark 1.6.0 Documentation. (2015). Retrieved from https://spark.apache.org/docs/1.6.0/programming-guide.html

Zhang, B., Cohen, W., Krackhardt, D., & Krishnan, R. (2011). Extracting Subpopulations From Large. *Social Networks*.

Chapter 7

# Network–Based Targeting:
## Big Data Application in Mobile Industry

**Chu (Ivy) Dang**
*The Chinese University of Hong Kong, China*

## ABSTRACT

*This chapter focuses on two kinds of targeting in mobile industry: to target churning customers and to target potential customers. These two targeting strategies are very important goals in Customer Relationship Management (CRM). In the first part of the chapter, the author reviews churn prediction models and its applications. In the second part of the chapter, traditional innovation diffusion models are reviewed and agent-based models are explained in detail. Customers in telecom industry are usually connected by large and complex networks. To understand how network effects and consumer behaviors – such as churning and adopting – interplays with each other is of great significance. Therefore, detailed examples are given to network-based targeting analysis.*

## INTRODUCTION

According to statistics from International Telecommunication Union (ITU), there will be more than 7 billion mobile cellular subscriptions by end 2015, corresponding to a penetration rate of 97%. In developed regions, penetration rates are much higher.

DOI: 10.4018/978-1-5225-1750-4.ch007

For example, mobile subscriber penetration rate in Hong Kong reached 232.2% in April 2015 (Office of the Communications Authority, 2015). Mobile Internet connection are getting cheaper and easier thanks to technological progress, infrastructure deployment, and dropping prices. Globally, mobile broadband penetration reaches 47% in 2015, a value that increased 12 times since 2007 (ITU, 2015). All these figures give evidence to one bold prediction "Mobile to overtake fixed Internet access by 2014" made by Mary Meeker, an analyst at Kleiner Perkins Caufield Byers (KPCB) who reviews technology trends annually. With such an increasing customer base and huge potential market, companies are facing great opportunities to enlarge their existing market and gain more profit. Yet, their competitors are also preparing themselves to share this big market. Fierce competition makes customer retention and customer acquisition more difficult than ever before.

Consumers have many choices over mobile service providers and mobile devices providers. For mobile service provider, to keep the existing customer from churning is the main focus of their CRM as customer churn often entails great loss to mobile service providers. Whereas, for mobile device providers, they often launch new handsets every year to attract new customers or allure existing customers to upgrade their devices. Therefore, mobile device providers focus more on customer acquisition. According to Recon Analytics (Entner, 2011), American people replace mobile phones every two years.

This chapter will focus on the first step in customer retention strategy and customer acquisition strategy: *targeting*. Targeting is of great importance in CRM. If targeting goes wrong, companies will waste resources and efforts in marketing practice, leading to great financial loss. With such large and diverse customer base, customer related data in mobile industry is often very big. In order to properly process and analyze these data, big data applications are necessary. One important big data application in mobile industry is targeting. This chapter will focus on such models and applications. Especially, network-based targeting analysis will be emphasized. The whole chapter will be arranged as follows. The next section will introduce the background of this chapter. Then the author reviews models and applications in two kinds of targeting: targeting churning customers and targeting potential customers. Lastly, the conclusion of this chapter is given.

## BACKGROUND

For any company, customer loyalty is an essential part for profit maximization. To obtain customer loyalty involves two steps: obtaining new costumers and then keep them from churning. The first step is called customer acquisition and the second is customer retention. Actually customer acquisition and customer retention is also the ultimate goal of most marketing strategies. For example, "free trial before buy" is a frequently used marketing strategy to attract new customers. "Membership" is also a common marketing strategy to keep existing customers. To obtain new customers and keep them loyal to the company is especially important for mobile industry in many countries since there are typically many wireless service providers as well as many mobile device providers within a region and customers can easily switch between them.

In Hong Kong, which has one of the highest mobile density in the world, for example, competition in public mobile services is vibrant. There are four main mobile network operators, namely, China Mobile Hong Kong Company Limited (CMHK), Hong Kong Telecommunications Limited (HKT), Hutchison Telephone Company Limited and SmarTone Mobile Communications Limited, providing a wide range of public mobile services (Hong Kong Government, 2015). Figure 1 shows the market share spectrum of mobile operators in Hong Kong. Indeed the competition within the mobile service sector is fierce. Operators use a variety of marketing strategies to maintain customer loyalty, ensuring high retention rate. One of the most frequently used strategy is Family Plan Packages, which give discount for communication between family members. Such plans also have dramatic consequence if customer churns. This will potentially induce their family members to churn because communication fee within the same operator is usually cheaper. As a result, customer churn will cause dramatic financial consequences for mobile operators. To tackle such churning effect, mobile service providers are developing sophisticated churn management strategies. First, they rank customers based on their estimated propensity to churn. Second, they offer retention incentives to a subset of customers at the top of the churn ranking. The core of this strategy is to *target* customer who are likely to churn. To be specific, such targeting strategy is usually called churn prediction. If churn prediction is inaccurate, companies will waste their money and effort on customers who would stay anyway. Therefore, companies want to predict churn as accurate as possible. The first part of this chapter will review the models and applications in targeting customers who are likely to churn. Specifically, the author will focus on models using big data technique and network analysis.

Besides mobile service provider, another important sector in mobile industry is mobile device provider. Figure 2 illustrates the market share of mobile phone of various brands in Hong Kong (by unit of shipment) in 2013. Although there are many mobile phone brands in the market, most of the market share is taken by two biggest players: Samsung and Apple. Each year, Samsung and Apple will announce its new mobile phone model and launch massive marketing campaign to allure customers buying new phones. According to Recon Analytics (Entner, 2011), on average American mobile phone user replace their mobile phones every 2 years. With such high replacement rate, customer retention is very hard to achieve. There-

*Figure 1. Mobile spectrum market share in Hong Kong*
Source: Office of the Communication Authority.

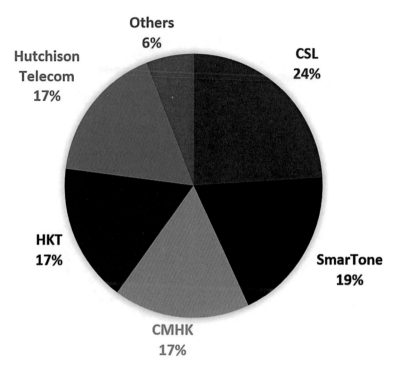

fore the marketing strategy of keeping customer loyalty for device providers is different from that of the mobile service providers. Mobile phone companies usually use new products to obtain new customers and allure existing customers to upgrade their handsets. That is to say, their marketing strategy focuses on customer acquisition instead of customer retention. From the companies' perspective, customer acquisition is equivalent to new product adoption. To understand how a new product diffuses in the market is essential for understanding the product adoption behavior of the customers. One of the most important steps in customer acquisition strategy is to *target* customers who are most likely to buy new products and then offer them incentives to induce buying behavior. The second part of this chapter will review the models and applications in targeting customers who are likely to adopt new products. The author will emphasis on models taking advantage of big data technique and network analysis.

*Figure 2. Analysis of market share by brands of mobile phone*
Source: Frost & Sullivan.

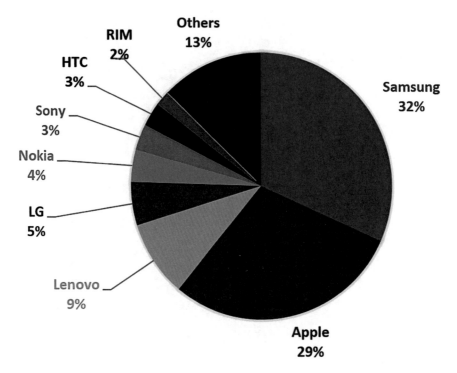

## TARGETING CHURNING CUSTOMERS

"Churn" is a word derived from change and turn. It means the discontinuation of a contract. Typically to gain new customers will cost much more than to retain the existing customers. Therefore, one of the most important goals in CRM is churn management. Churn management is about two things: predicting *who* is most likely to churn and analyzing *why* they churn. The "who" question helps mobile service providers target the right customers. The "why" question helps companies design proper customer retention strategies for these customers. If companies only predict churning without analyzing the underlying driving force of such behavior, retention effort will eventually be wasted. Churn prediction answers the "who" question and churn analysis answers the "why" question. In this section, the author will explain churn management from such two aspects: churn prediction and churn analysis.

In mobile industry, customer related data set are typically very big. Such data contains many information about customers, such as call details, 3G/4G data usage details, contractual information, demographics etc. With increasing popularity of mobile service and increasing frequency of use (ITU, 2015), such data set is becoming bigger and bigger. Therefore, CRM in mobile industry depends heavily on big data applications. Particularly, data mining technique is most frequently used (Wei & Chiu, 2002).

Data mining refers to techniques which analyze a very large data set from various dimensions and then "dig out" previously unknown, non-trivial, consistent patterns and/or systematic relationships between variables (Berry & Linoff, 2004; Chen, Han, & Yu, 1996). Such pattern and/or relationships can then be used to predict future events. The "mining" process typically includes a learning algorithm such that the model can be adaptive to different inputs. In CRM, the most commonly used data mining techniques include clustering (Ngai, Xiu, & Chau, 2009), classification (Neslin, Gupta, Kamakura, Lu, &Mason, 2006; Lemmens & Croux, 2006), genetic models (Eiben, Koudijs, & Slisser, 1998), neural network (Tsai & Lu, 2009) etc. Since churn management simply cares whether customers will churn or not, such a binary behavior with known categories is best captured by classification models.

Classification, in simple words, is assigning data to one of some pre-defined categories. Usually, one classification model is composed of three steps. First, classify training data into some predefined categories, defined as label set. The features of the training data compose the feature set. Second, employ a learning algorithm to identify a model that best fit the relationship between the feature set and the label set. Informally, this classification model can be called a target function $f$. A target function maps each feature set $x$ to one of the predefined labels $y$. Third, use classification model to classify test data with unknown class labels into different categories according to their feature set. After that, evaluate the model based on the accuracy of their prediction of the test data. Figure 3 illustrates how classification model works. A real life example would be an email system which can classify an email into either spam or not spam. In churn prediction model, such classification model would be the one which can classify a subscriber into churn or not churn according to their features such as demographics, contractual data, customer service logs, call details, complaint data, bill and payment information etc. Classification models applied in churn prediction include Naïve Bayes classifier, decision tree classifier, decision rule classifier, neural network etc.

## General Customer Churn Management Methodology

1.  **Obtain Data Set:** Mobile service provider needs to extract huge volume of customer related data. These data can be grouped into four categories (Van

*Figure 3. How classification model works*

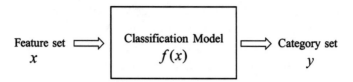

den Poel & Lariviere, 2004): customer behaviors (calling details, contractual information etc.), customer perceptions (can be measured via survey), customer demographics (age, gender, education etc.), and macro-environment variables (natural disaster, political revolution etc.).

2.  **Data Reduction and Cleaning:** Typically the first-hand data set is very large with many noises. Data reduction will help to ignore unimportant variables and features such that the data can be reduced into manageable size. Data cleaning includes eliminating noisy data and estimating missing values. Interpolation is the most widely used method for missing value estimation.

3.  **Build Model:** Randomly use a training set of data which contains both churners and non-churners to build churn prediction model. Typically this process is composed of a classification model with a learning algorithm. The churn prediction model will be reviewed in the following section.

4.  **Evaluate Model:** After construction, test data with known category (churn or not churn) will be used in evaluation. Quality measures in churn prediction include *accuracy*, *sensitivity* and *specifity*. The most import quality measure in the evaluation process is *accuracy*. Accuracy is defined as the percentage of correct predictions. These measures can be obtained using a confusion matrix. Figure 4 is a confusion matrix in churn prediction models. According to Domingos (1999), an accuracy of about 90% is sufficient for a model to provide satisfactory churn prediction.

5.  **Adjust Model:** If the model is not satisfactory after evaluation, it needs to be adjusted until it gets good quality measure results, such as a 90% accuracy score.

6.  **Predict Churn:** Now the model is ready to predict customers who are likely to churn given a new data set with known features but unknown category.

7.  **Analyze Reasons:** A huge drawback of churn prediction model is that it cannot answer the "why" question: why do those people want to churn? Effective customer retention strategies rely heavily on these reasons since the time and resources for retention is limited (Lazarov & Capota, 2007). Churn analysis models are used to analyze these reasons. Some of these reasons include social learning and network effect (Hu, Yang & Xu, 2015).

*Table 1. Confusion matrix*

|  | Actual Churners | Actual Non-Churners |
|---|---|---|
| Predicted Churners | $a_{11}$ | $a_{12}$ |
| Predicted Non-Churners | $a_{21}$ | $a_{22}$ |

8. **Design Customer Retention Strategies:** After analyzing the reasons behind churn behavior, proper customer retention strategies can be designed. For example, if network effect dominates, strategies like family plan packages or developing exclusive network games would be effective. However, if social learning dominates, strategies like publicly recognizable handsets or accessories and positive word-of-mouth would be better (Hu et al., 2015).

Step 1 to 6 is churn prediction procedures and step 7 to 8 is churn analysis procedure. In the following section, the author will briefly review these models and their applications in mobile churn management.

## Churn Prediction Models

Neslin et al. (2006) conducts a tournament in which both academics and practitioners use churn prediction models. In this practice, various churn prediction models are introduced. Among them, 45% are logistic regression, 23% are decision trees, and 11% are neutral networks. Other methods include discriminant analysis (9%), cluster analysis (7%), and Bayes method (5%). The author will review these models in this section.

## Logit Model

Before big data age, the simplest and most traditional way of churn predicting is logit model (Hosmer & Lemeshow, 2004). Logit model, also called logistic regression, is frequently used to model binary outcome variables. It can be seen as a binary classification model. In the logit model the log odds of the outcome is modeled as a linear combination of the predictor variables. In churn prediction, the binary dependent variable is simply $y_i = churn$ or $y_i = not\ churn$. Vector $\boldsymbol{x} = \left( x_{i1}, x_{i2}, \ldots, x_{in} \right)$ contains features used for churn prediction which includes customer demographics, contractual data, customer service logs, call details, complaint data, bill and payment information etc. Regression coefficients $\beta = (\beta_0, \beta_1, \beta_2, \ldots)$ can be estimated by maximum likelihood estimation (MLE) using previous customer information (training data). After obtaining the coefficients, Equation 1 is used for calculating the probability of customer churn:

$$prob\left(y_i = churn\right) = \frac{Exp\left(\beta_0 + \sum_{k=1}^{n}\beta_k x_{ik}\right)}{1 + Exp\left(\beta_0 + \sum_{k=1}^{n}\beta_k x_{ik}\right)} \tag{1}$$

Logit model often serves as the benchmark model when comparing the performance of different churn prediction models (Neslin et al., 2006; Lemmenaswb s & Croux, 2006). Neslin et al. (2006) gives a comprehensive comparison of five most common churn prediction techniques to investigate their predictive accuracy. Lemmens and Croux (2006) apply bagging and boosting technique to churn prediction models. After comparing it with the standard logit model, they find significant improved prediction power in term of Gini coefficient top-decile lift. Logistic regression can also be used in feature selection (Masand, Datta, Mani, & Li, 1999). Usually, the data warehouse collects many features of each customer. To select the most predictive features as the input for churn prediction is of significant importance.

## Decision Trees (DT)

As one of the most frequently used data mining methods (Berry & Linoff, 2004), DT is a powerful tool for classification and predition by finding out the patterns or relationships between data. DT is made up in the form of a tree built by making child-nodes until each branch reaches the terminal node. There are three kinds of nodes in a decision tree: root node, internal nodes, and leaf or terminal nodes. Root node and internal nodes contain feature test conditions to separate observations that have different characteristics. Each leaf node is assigned a class label. Figure 4 is a simple decision tree to judge whether a subscriber will churn or not. The classification starts from the root node with one feature test condition such as if the age of the subscriber is below sixty. If yes, the next internal node will judge whether the duration of his/her usual call is below 2 minutes. If yes, the subscriber will be classified as churner. If not, he/she will be classified as non-churner. Based on different split criterion, some of the algorithms for DT are CHAID (Kass, 1980), CART (Breiman, Friedman, Stone, & Olshen, 1984), and C4.5 (Quinlan, 2014).

Many researchers utilize decision tree classifier to model costumer churn behavior (Breiman et al., 1984; Wei & Chiu, 2002; Bin, Peiji, & Juan, 2007; Hung, Yen, & Wang, 2006). For example, based on customer demographics and contractual data, Breiman et al. (1984) use CART algorithm to predict customer churn.

*Figure 4. Decision tree for churn prediction*

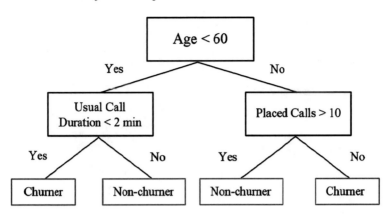

However, not all companies have customer demographic information. To tackle the problem of unavailability of customer demographics in the mobile service provider investigated, Wei and Chiu (2002) design a decision tree model to identify potential churners at contract levels. Their empirical evaluation results suggest that when more recent call details are employed for model construct, the prediction effectiveness will be more satisfactory. Also aiming at overcoming the limitations of lack of information of customers of Personal Handyphone System Service (PHSS), Bin et al. (2007) improve the existing decision tree model from three aspects: changing sub-periods for training data sets, changing misclassification cost in churn model, and changing sample methods for training data sets. By carrying out three research experiments, some optimal parameters of models are found. Hung et al. (2006) empirically explore the effectiveness of data mining technique in churn prediction with a customer related data set provided by a wireless telecom company in Taiwan. Particularly, the data mining technique used by Hung et al. (2006) are decision tree model and neutral network. They show that both models can deliver accurate churn prediction by using customer demographics, billing information, contract/service status, call detail records, and service change log.

## Neutral Network

The statistical technique neutral network is inspired by the way human brains process information. This information processing system is composed of a large number of highly interconnected "neurons" (processing elements) which send messages to each other. These elements work in unison to solve specific problems. By "learning" from previous experience, the weights put on these connections can be tuned, making this system adaptive to inputs. Such data mining technique is wide-

*Figure 5. Neutral network which is equivalent to linear regression*

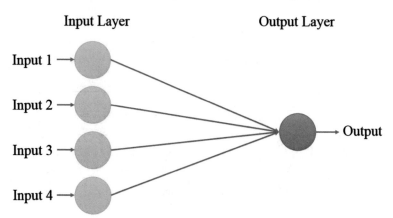

ly used in computer vision and voice recognition. Neutral network classifier is essentially a function $f : x \rightarrow y$ which maps feature set (inputs) to a label set (outputs). This function typically incorporates a learning rule such that it can be adaptive learning. In addition, this function allows complex nonlinear relationships between the feature set and label set. Figure 6 is one of the simplest neutral network which is equivalent to linear regression. There are four inputs or features in this model with attached "weights" for each of them. The prediction function is a linear combination of these four inputs. The weights are selected using a learning algorithm such that it can minimize a cost function, such as mean squared error (MSE).

Tsai and Lu (2009) evaluate the performance of two churn prediction models utilizing two hybrid neutral network techniques. Particularly, these two techniques are back-propagation artificial neural networks (ANN) and self-organizing maps (SOM). The hybrid models are ANN combined with ANN (ANN + ANN) and SOM combined with ANN (SOM + ANN). Their experiments show that the two hybrid models perform better than the benchmark neutral network model in terms of prediction accuracy and Types I and II errors. There are general two downsides of neural network analysis in churn prediction. First, because neural network technique is a holistic approach to learning by encoding the classification model in the weights between nodes, its resulting knowledge often lacks interpretability. Second, due to the iterative nature of neural network analysis, it requires a long training time which will be a big problem for very large data set.

## Naïve Bayes Classifier

Frequentist statistics defines an event's probability as the limit of its relative frequency in a large number of trials. However Bayesian probabilities is expressed in terms of level of certainty relating to a potential outcome. The simplest way to incorporate Bayesian statistics into churn prediction is Naïve Bayes classifier (Lazarov, Iba, & Thompson, 1992). Naïve Bayes is a type of supervised-learning module that contains examples of the input-target mapping which the model tries to learn. Such models make predictions about new data based on previous experience. The learning process is done by updating prior belief by posterior belief.

Given the same model, Naive Bayes calculates the probability that a given input data (e.g., the input feature vector $\boldsymbol{x}_i = (x_{i1}, x_{i2}, \ldots, x_{in})$) in churn prediction case) is generated by certain data generation process (DGP). Denote $\theta \in \Theta$ as the possible parameter vector in DGP. The posterior belief can be written as:

$$p\left(\theta \mid \boldsymbol{x}_i\right) = p\left(\boldsymbol{x}_i \mid \theta\right) p\left(\theta\right) = p\left(x_{i1}, x_{i2}, \ldots, x_{in} \mid \theta\right) p\left(\theta\right)$$

where $p\left(\theta\right)$ is the prior belief of $\theta$. Naive Bayes assumes that the conditional probabilities of the independent variables are statistically independent. So the above equation can be re-written as:

$$p\left(\theta \mid \boldsymbol{x}_i\right) = p\left(\theta\right) \prod_{k=1}^{n} p\left(x_{ik} \mid \theta\right).$$

The estimation criteria of $\theta$ is given by:

$$\hat{\theta} = \arg\max_{\theta \in \Theta} p\left(\theta \mid \boldsymbol{x}_i\right).$$

Based on an Oracle database of fifty thousand real customers, Nath and Behara (2003) build a working database system for customer churn prediction using Naïve Bayes classifier. Their model obtained 68% predictive accuracy. Another application of Bayesian method is Kisioglu and Topcu (2011) who apply Bayesian belief network approach to identify potential churners and find out the most important features that influence customer churning behavior in mobile industry.

## Churn Analysis Models

A huge drawback of churn prediction model is that the driving force behind such churning behavior is unknown. In order to understand such driving force, models which not only predict churning but also give insights on the reasons behind churning are constructed. Some of these techniques include evolutionary learning and self-organizing maps.

### Evolutionary Learning

Au, Chan, & Yao (2003) propose a new data mining algorithm, called data mining by evolutionary learning (DMEL), to predict and analyze churn behavior. Unlike most classification models which fail to give likelihood for each prediction, DMEL can estimate the accuracy of each prediction. The model can be summarized as follows: 1. Using genetic algorithm to construct rules iteratively; 2. Identify interesting rules and measure them using objective interestingness measure; 3. Calculate the probability that the attribute values of an observation can be correctly determined using the encoded rules; 4. Estimate the likelihood of each prediction. Using the likelihood score and customer feature set, further analysis of customer churning is possible. For example, customers can be grouped into those with high likelihood and those with low likelihood. By identifying groups and features, companies will gain insight on real reasons for churning.

### Self-Organizing Maps (SOM)

SOM is one of the clustering models in unsupervised learning. Clustering refers to a method which partitions a set of patterns into clusters without predefined classes. Cluster analysis refers to the grouping of a set of data object into clusters. In particular, no predefined classes are assigned (Jain, Murty, & Flyn, 1999). SOM was proposed by Kohonen (1987) and proved extremely useful when the input data are with high dimensionality and complexity. SOM is used to discover relationships in a dataset and then cluster data according to their similarity. Ultsch (2002) utilizes a combination of emergent self-organizing maps, U-matrix methods and knowledge conversion technique to predict potential mobile phone churners. The output rules produced by the model help mobile service providers get better understanding of who the clients are, how profitable they are and why churning is happening. With such information, better retention strategies, such as marketing campaign and service upgrading can be designed.

## Network-Based Churn Analysis

As already been illustrated in the previous section, to understand what triggers consumers to switch wireless carriers is of great importance. Revealing the underlying mechanism behind this behavior will help telecommunication companies enhance product design and marketing strategy to avoid costumer churn, and ultimately optimize their customer retention efforts. For example, Dasgupta et al. (2008) examine the communication patterns of millions to address the role of social ties in the formation and growth of groups in a mobile network. They show that customers are more likely to switch wireless carriers if more of their contacts from the same operator switch. With the availability of data about individual interactions, there is a growing interest in understanding the mechanisms regarding this influence (Peres, Muller, & Mahajan, 2010).

Mantian Hu and her co-authors (Hu, Yang, & Xu, 2015) successfully identified the two underlying behavioral mechanisms, social learning and network effects, to explain such assimilation effect by proposing a two-step dynamic forward looking model. Figure 6 illustrates the modelling framework. According to their study, the individual costumer decides whether or not to switch carrier according to three sources of information:

1.  Their own user experience of the current carrier and their idiosyncratic belief on the alternative carrier;
2.  Feedbacks from their contacts who have switched to update their own quality expectations and learn about the alternative carriers;
3.  Network effect induced by the switching decisions (for example, the more contacts a customer has, the more benefit she potentially receives).

(2) and (3) are two fundamentally different mechanisms why indiciduals imitates one another: the former is based on information exchange and it is called *social learning* (Moretti, 2011); the latter is based on direct benefit from aligning their behavior to others and it is called *network effect* (Katz & Shapiro, 1985). The network effect is a newly raised topic and it differs from social learning since others' behaviors are affecting someone's payoff directly, rather than indirectly by changing his information set. By proposing a dynamic structural model with strategic interpersonal interactions, Hu et al. (2015) find strong evidence for both social learning and network effect in mobile customer churn behavior. They show by simulation that 1% change in network size will lead to 11.5% change in customer retention rate in the same direction. And two-thirds of such impact can be attributed to network effects and one-third to social learning effects. Apart from simulation, they also apply this framework to data from a wireless carrier in a European country. Their

*Figure 6. Modelling framework*
Source: Hu et al. (2015).

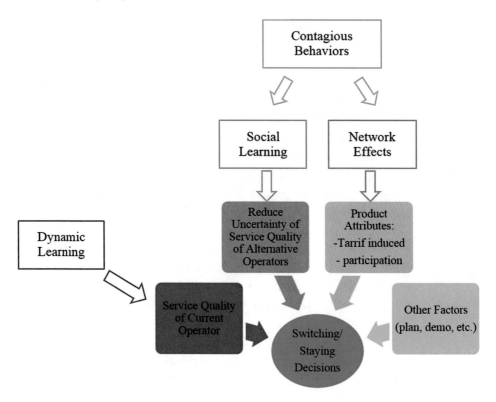

empirical results indicate that, after controlling for mobile plan details and demographic information, learning from own usage experience, learning from contact neighbors' decisions, and network effects all have significant effects on customer mobile service switching decisions.

This is the first study to incorporate a learning framework within a social network context, where people's decisions affect one another endogenously. It disentangles two important mechanisms, social learning and network effects, by explicitly modeling the direct utility impact and the learning process. The disentanglement of the two behavioral mechanisms will provide great insight on how to avoid costumer churn or how to attract new customers for telecom companies. Given the fact that telecom industry experiences an average of 30-35 percent annual churn rate and it costs 5-10 times more to recruit a new customer than to retain an existing one, customer retention has now become even more important than customer acquisition for mobile service providers. Hu et al. (2015) suggest that if network effect plays a dominant role, companies should focus on maintaining a large customer base to keep existing ones and at the same time attract new customers. However, if social learning prevails, information exchange is the key. So strategies like publicly recognizable handsets or accessories and positive word-of-mouth will be better. Moreover, this research also provides another angle to estimate wireless carrier's long-term customer retention rate and its profitability.

## TARGETING POTENTIAL CUSTOMERS

Besides targeting customers who are most likely to churn, another important targeting goal is potential customers. In CRM, the ultimate goal of customer acquisition management is the acquisition of potential customers in an effective fashion (Saylor, 2004). "Effective" here means using the least effort and spending the least amount of money to acquire as many new customers as possible. Therefore, there are two main steps in customer acquisition management: first, target the right people; second, develop the right acquisition strategies. These strategies include customer loyalty programs, joining charitable events, launch mass marketing campaigns etc. In mobile industry, as already has been illustrated, the mobile device providers, such as mobile phone companies, new customer acquisition is more crucial than existing customer retention. Data shows (Entner, 2011) that typically one changes mobile phone every two years in countries like the United States. This is probably due to the low switching cost and fast technology innovation within the mobile device sector. According to Moore's Law, processor speeds, or overall processing power for computers will double every two years. Moreover, researchers showed that quality and inflation adjusted price of IT equipment declined 16% per year on average over five decades from 1959 to 2009 (Nambiar & Poess, 2011). More advanced technology with lower prices, this gives another reason for customers to adopt new handsets. In mobile device sector, the competition is even fiercer than the mobile service sector. To maintain or/and enlarge market share, mobile phone companies for example, develop many categories of phones to meet the needs of different customers. Particularly, big companies like Apple, Samsung, and Xiaomi launch their featured product every year to attract new customers and allure existing customers to upgrade their devices. In the meantime, they face extremely fierce competition with each other. A successful new product will diffuse very fast and take over competitor's market. For example, Samsung launched Galaxy S3 and Note II in 2012 which was a big success over Apple's iPhone 5. Apple's global smartphone market share dropped from 23% in the first quarter of 2012 to only about 12% in the third quarter of 2014 (International Data Corporation, 2015). However, after Apple launched its iPhone 6 model which was loved by the consumers in September 2014, its market share quickly climbed to the same level of Samsung by the end of 2014

(International Data Corporation, 2015). Customer loyalty in mobile phone sector is especially low, so companies devote many efforts to acquire new customers.

From companies' perspective, before they come up with any marketing strategies, they must first understand *who* adopt their product, *how* product diffuse in the marketplace and *what* drives diffusion process. The "who", "how" and "what" question can be captured in the new product diffusion process, which can be explained in innovation diffusion models or new product diffusion models. For traditional commercial products such as foods and durable goods, companies usually are incapable of obtaining customer related data and their social connections. However, one big advantage in mobile industry is the easy acquisition of consumer's behavioral data (e.g., detailed calling information), demographic data (age, sex etc.) and their mobile device information. With such rich data, companies can investigate the mobile device diffusion process conveniently. Diffusion denotes the spread of an innovation in the market (Peres et al., 2010). Rogers (2010) defines diffusion of innovation as "the process by which an innovation is communicated over time among the members of a social system. It is a special type of communication, in that the messages are concerned with new ideas". The diffusion process has many managerial implications. Some of those include how market mix (price, place, promotion, and product) affect the adoption process and brand image, how competitions influence the growth pattern of products etc.

## Innovation Diffusion Theory

The innovation diffusion theory was first introduced by Rogers (1962). It seeks to understand how new ideas, products and practices spread throughout a society over time. In Rogers' model, the aggregate adopters follows an S-shaped curve. According to the different positions on the S-shaped curve, the adopters are grouped into five different categories: innovators, early adopters, early majority, late majority, and laggards. However Rogers' model is a conceptual framework rather than a quantitative model. Bass (1969) introduced a quantitative diffusion model which is the most frequently used model for diffusion research. Before Bass model, there are two other quantitative models proposed by Fourt and Woodlock (1960) and Mansfield (1961). Fourt and Woodlock (1960) is a pure innovative model where they assume new product adopters are influenced by external information such as advertising and mass marketing campaign, rather than internal influence such as word-of-mouth. On the contrary, Mansfield (1961) constructs a pure imitative model by assuming people adopt new things by internal influences such as word-of-mouth and interpersonal interactions. Bass model combines these two effects together. The Bass model assumes both external information (e.g., mass media) and internal information (e.g., word-of-mouth) influence potential adopters of innovation. Bass model has very good predictive power and it is the theoretical foundation for many diffusion models. These models are all traditional innovation diffusion models where they use aggregate level data. Therefore, they understand diffusion process from a macro-level.

In order to fully investigate the driving forces for each individual, micro-level diffusion theories are necessary. In the review of Mahajan, Muller, and Bass (1990), the authors also call for individual-level based diffusion models to explore the social communication pattern, and its impact on product perceptions, preferences and ultimate adoption decision. Such micro-level diffusion models will also reveal the relationships between individual adoptions and aggregate growth. These micro-level models include: microeconomic models, stochastic choice models and agent-based models (ABM). Microeconomic models and stochastic choice models typically rely on strong distributional assumptions about individual information and they also lack analysis of aggregated variables. Consumer-level data in mobile industry is very rich and thanks to the advanced information technology, especially to those related to big data application, researchers and companies are worrying less about long computation time and small computational power of computers. Therefore, in mobile industry, agent-based models are the most popular individual-level diffusion models. Many agent-based models are either extensions or variations of the Bass model with the same underlying logic. Cellular automata is one of these models. Agent-based models have many advantages, for example, they can incorporate heterogeneity and dynamics into individual behaviors, and they also link disaggregate individual-level choices to aggregate macro-level variables.

In the Bass model, the decision to adopt a new product at the population level is modeled as a hazard rate without considering the local network effects. However, in mobile industry, people are connected and communicated through their mobile devices. Ignoring the heterogeneity of consumers' local networks is one of the biggest drawbacks of the Bass model. To tackle this problem, agent-based models with network structure analysis are developed. For example, Hu, Hsieh and Jia (2014) investigated how dynamic network structure affects consumer's new smartphone adoption behavior. Their research showed that network structure is a good predictor of social influence. This study will be introduced by the author later in detail.

In the following section, the author will first briefly introduce the traditional Bass model and its agent-based extensions. Then a combination of agent-based model and network structure analysis will be introduced by a case study.

## The Bass Models

The work done by Fourt and Woodlock (1960), Mansfield (1961) and Bass (1969) contribute to the foundation of traditional parsimonious empirical diffusion models, among which Bass (1969) is the most popular model. In the diffusion literature, there are many models which are based on refinement or extension of Bass model with the basic premise of the model (Mahajan et al., 1990). Therefore the author will briefly go through these three models.

## Innovative Models

After examining the marketing penetration curves of many new products, Fourt and Woodlock (1960) proposed an innovation diffusion model which assumes adoption behavior is influenced by external sources such as advertising and marketing campaigns. In addition, the cumulative sales curve follows an exponential shape. The mathematic formula is:

$$f_t = \Delta Q_t / Q_p = rM \left(1 - r\right)^{t-1}$$

where $\Delta Q_t$ denotes the change in cumulative product sales at time $t$ and $Q_p$ denotes potential sales. $r$ is the rate of penetration of potential sales and $M$ is the ratio of total potential sales to all buyers. Parameters $Q_p$, $r$ and $M$ are constant across time.

## Imitative Models

Imitative models (Mansfield, 1961; Fisher & Pry, 1972) assume adoptions are mainly influenced by internal sources such as word-of-mouth and interpersonal interactions. For example, Fisher and Pry (1972) assume that new product adoption rate is dependent on the fraction of the old product which are still in use. They regard advancing technology as a set of substitution processes. Therefore new products act as substitutions to old ones. The new product adoption rate in Fisher and Pry (1972) model is the classic logistic S-shaped curve:

$$f_t = \frac{1}{1 + e^{-2\alpha\left(t - t_0\right)}}$$

where $\alpha$ is half the annual fractional growth in the early years. $\alpha$ and $t_0$ is the time at which $f_t = 1 / 2$. The S-shaped curve is characterized by two constants: the early growth rate $\alpha$ and the time at which the substitution is half complete $t_0$.

## Bass Model: A Combination

The Bass model assumes that new product adopters are influenced by two types of communication: mass media (external source) and interpersonal communication (internal source). Therefore external sources will have greater impact on innovative customers who is the main driving force for product takeoff. Whereas internal sources will impact more on imitative customers who are more likely to buy if more of their friends are buying. There imitative customer contribute more during the later periods of diffusion process (Rogers, 2010). One of the basic assumptions in Bass model is that the probability of adopting a new product at time $t$ is a linear function of the number of previous adopters. In mathematic term, that is

$$f_t = \left(p + qF_t\right)\left(1 - F_t\right) \tag{2}$$

where $f_t$ is the likelihood of purchase at time $t$ and $F_t$ is the cumulative fraction of adopters at time $t$. Parameter $p$ captures the innovative influence and $q$ captures the imitative influence. Aggregate models are primarily concerned with modeling $\Delta Q_t$, the flow of consumers from potential market $m$ to current market (Mahajan & Muller, 1979). Equation 3 is typically used to calculate $\Delta Q_t$.

$$\Delta Q_t = \left(p + qQ_t / m\right)\left(m - Q_t\right) \tag{3}$$

$Q_t$ is the cumulative adoption and $m$ is the market size. $\Delta Q_t$ is a bell-shaped curve showing the speed of diffusion. $Q_t$ is a S-shaped curve which shows diffusion in a cumulative way.

Some of the advantages of Bass model are: first, it gives an analytically tractable way to interpret the whole market diffusion process; second, Bass model use more readily available market-level data to forecast sales; lastly, the estimation methods for Bass model are well documented (Bass, 1969; Schmittlein & Mahajan, 1982; Srinivasan & Mason, 1986). Despite these advantages, Bass model unavoidably exhibits many drawbacks. For example, it doesn't incorporate population hetero-geneity and dynamics into the model. In addition, it ignores the structure of social interactions. Moreover, the linkage between micro-level adoption behavior and macro-level growth is missed. In order to conquer these limitations, agent-based models are called in.

## Agent-Based Bass Model

Bass model can be easily formalized using agent-based model. Therefore, Bass model is a special case that can also be captured by an analogous agent-based model (Kiesling, Günther, Stummer, & Wakolbinger, 2012). The author will explain this process to illustrate the differences in modeling. Assume there are $M$ agents indexed by $i = 1, 2, \ldots, M$. Each of them is in one of the two states: potential adopter ($x_i = 0$) or adopter ($x_i = 1$). In the original Bass model, probability of adopting a new product at time $t$ is a linear function of the number of previous adopters which is expressed using Equation 2. Equation 4 shows the agent-based analogy. Each agent's individual probability of adoption is influenced uniformly by the adoption state of all other agents. In addition, global connectedness and homogeneity are also implied in the equation.

$$
f_i = \left( p + \frac{\sum_{i=1}^{M} x_i}{M} q \right) \left( 1 - x_i \right) \tag{4}
$$

Algorithms such as discrete time updating process are adopted to do the simulation. Since stochastic process is often included in the algorithm, ABM typically does not provide a single analytical solution, but involves uncertainty and variability. The relationship between ABMs and Bass model was studies by, for example, Fibich and Gibori (2010), and Rahmandad and Sterman (2008). Fibich and Gibori (2010) show that Bass model provide an upper bound for the aggregate diffusion dynamics in agent-based models with "any" spatial structure. Rahmandad and Sterman (2008) compare ABMs with differential equation (DE) models and examine the impact of individual heterogeneity and different network topologies.

## Cellular Automata

The cellular automata model was originally introduced by Stan Ulam and John von Neumann in the 1940s to provide a formal framework for investigating the behavior of complex, extended systems (von Neumann & Burks, 1966). Agent-based cellular automata models differ from the traditional model in the following ways: 1. the state of an agent is more complex; 2. local interaction structures can be heterogeneous. Take a simple cellular automata model which fulfills the assumptions in basic Bass model for example (Goldenberg, Libai, & Muller, 2001). The cells in this model are all potential buyers and they all interact with each other, with binary state value "0" or "1". State "0" represents potential buyers who do not adopt innovation. State

"1" represents potential buyers who adopt the innovation. Like Bass model, the mechanisms that govern the transitions of potential buyers are external influences such as mass media and internal influences such as word-of-mouth. The former is captured by parameter $p_i$ and the latter is captured by parameter $q_i$. Equation 5 illustrates the mathematical formula. $f_i(t)$ is the probability of adoption at time $t$ and $Q_i(t)$ the cumulative number of adopters at time $t$. In this model, $p_i$ and $q_i$ capture the heterogeneity of consumers in terms of external and internal influences, which differs from Bass model where the influences are homogeneous.

$$f_i(t) = 1 - (1 - p_i)(1 - q_i)^{Q_i(t)} \tag{5}$$

## Network-Based Diffusion Analysis

Disaggregate models such as ABMs has many desirable properties. Some of them include: 1. It captures interpersonal interactions; 2. Heterogeneity can be incorporated into the model; 3. The model can include network dynamics and structures without knowing the exact global interdependencies (Borshchev & Fillipov, 2004); 4. Aggregate level variables can be obtained from the micro-level individual behaviors; 5. Linkage between micro-level adoption behavior and macro-level growth is recovered. Since agent-based models build on decisions of each individual, they are therefore more behaviorally based. In a word, agent-based models aim to capture macro-level phenomena such as diffusion process by simulating the decision process of each individual. Mobile industry therefore is a perfect place to apply these techniques.

Interpersonal communication is seen as a key influence on new product diffusion process (Rogers, 2010; Mahajan et al., 1990). In the agent-based model, interpersonal communication channels can be represented by networks. Usually, such network-level influence is called social influence. The notion of social influence in the diffusion of new products has been well accepted (Iyengar, Bulte, & Lee, 2015). In mobile industry, such influence is especially pronounced since people are connected by their wireless devices. For mobile device companies, in order to target the right potential buyers, they must first understand the relationship between social influence and new product diffusion process. While many studies have been focusing on identifying and quantifying the social influence effect (Van den Bulte & Lilien, 2001; Manchanda, Xie, & Youn, 2008; Iyengar, Van den Bulte & Valente, 2011), little has been done to predict its occurrence and even little has been done to reveal how network structures affect the adoption behavior.

By using a model which captures the co-evolution of both network structures and product diffusion process, Hu et al. (2014) reveals how dynamic network structures affect consumers' new smartphone adoption decisions. They get access to the complete information of the entire customer base of a major Chinese wireless carrier in two medium-sized cities in western China. Based on the calling records, they extract individual social network using snowball sampling. Since new cellphone diffusion process in each individual networks are different, they propose to examine how network topology interplays with the effect of social influence. Particularly, they examine this using the diffusion of Samsung smart phones. A difficult task of this study is how to distinguish social influence effect from homophily effect. Hu et al. (2014) solve this by using the stochastic agent-based dynamic network formation model. They use this model to identify and measure the sizes of homophily effect on network formation and social influence. The stochastic agent-based dynamic network formation model (Snijders, 1996, 2001; Snijders, Steglich, & Schweinberger, 2007) has been widely applied to model co-evolution of a dynamic social network and diffusion of behaviors. However, Hu et al. (2014) is the first study to apply this model in marketing context. After estimating social influence, they use meta-analysis to link it with network structure measures. For example, they find network size, density, clustering coefficients, diversity, assortativity, epidemic threshold and position of initial adopters are good predictors of social influence.

Besides the relationship between network structure and social influence, Hu et al. (2014)'s study also sheds light on many other aspect of mobile industry. For example, they apply the analysis to three choice levels for Samsung smartphones: mobile phone level (Samsung Note II), higher brand-tier level (Samsung high-end phones), and brand level (Samsung phones). They find that 6.0% networks exhibit social influence for Samsung Note II adoption, 12.3% for Samsung high-end phone adoption, and 10.2% for Samsung brand adoption. This result is quite counterintuitive given the fact that the adoption rate for Samsung branded phones is higher than that for the Samsung high-end phones. Hu and her co-authors find significant and positive social influence effects in all three behavioral cases. Their study also empirically demonstrates that if all the friends of an individual adopt a product, then social influence increases the chances of that individual adopting the same product by 7.38 times. Moreover, the pervasive homophily effect driven by product adoption has not been observed in their study; individuals do not connect with one another just because they have the same phone.

Hu et al. (2014)'s study gives many insights for mobile device providers. For example, before companies target the right individuals, they should first target the right networks. In addition, adoption rate and social influence are two separate concepts and they do not necessarily correlate with each other. Lastly, companies should develop some network-level customer acquisition strategies before targeting the individual customers.

## CONCLUSION

In this chapter, the author gives a review of network-based targeting techniques for mobile industry. Emphasis is given to models which utilize big data applications. The author opens the chapter by explaining the concept of targeting and its importance. In mobile industry, there are two targeting goals: to target customers who are likely to churn and to target customers who are likely to adopt a new product. The main body of this chapter reviews churn prediction models and product diffusion models separately. Churn prediction models are used for targeting churning customers for mobile service providers. Product diffusion models are used to target potential buyers for mobile device providers. The network-based analysis of such targeting process is emphasized by the author. The author uses Hu et al. (2015) to explain network-based churn analysis. Network-based diffusion analysis is illustrated by introducing the work of Hu et al. (2014). Both of the two studies incorporate network-level analysis to investigate consumer behaviors.

The future research direction in mobile industry targeting is to refine the micro-level agent-based models and incorporate network-level analysis into them. Furthermore, big data techniques such as machine learning can be utilized to predict consumer behaviors. Lastly, developments in behavioral economics and psychology can be borrowed to analyze the decision making process of the consumers.

## REFERENCES

Au, W. H., Chan, K. C., & Yao, X. (2003). A novel evolutionary data mining algorithm with applications to churn prediction. *IEEE Transactions on Evolutionary Computation*, *7*(6), 532–545. doi:10.1109/TEVC.2003.819264

Bass, F. M. (1969). A new product growth model for consumer durables. *Management Science*, *15*(1), 215–227. doi:10.1287/mnsc.15.5.215

Berry, M. J., & Linoff, G. S. (2004). *Data mining techniques: for marketing, sales, and customer relationship management*. John Wiley & Sons.

Bin, L., Peiji, S., & Juan, L. (2007, June). Customer churn prediction based on the decision tree in personal handyphone system service. In *Service Systems and Service Management, 2007 International Conference on* (pp. 1-5). IEEE. doi:10.1109/ICSSSM.2007.4280145

Borshchev, A., & Filippov, A. (2004, July). From system dynamics and discrete event to practical agent based modeling: reasons, techniques, tools.*Proceedings of the 22nd international conference of the system dynamics society*, 22.

Breiman, L., Friedman, J., Stone, C. J., & Olshen, R. A. (1984). *Classification and regression trees*. CRC Press.

Cantono, S., & Silverberg, G. (2009). A percolation model of eco-innovation diffusion: The relationship between diffusion, learning economies and subsidies. *Technological Forecasting and Social Change, 76*(4), 487–496. doi:10.1016/j.techfore.2008.04.010

Chen, M. S., Han, J., & Yu, P. S. (1996). Data mining: An overview from a database perspective. *Knowledge and data Engineering. IEEE Transactions on, 8*(6), 866–883.

Dasgupta, K., Singh, R., Viswanathan, B., Chakraborty, D., Mukherjea, S., Nanavati, A. A., & Joshi, A. (2008, March). Social ties and their relevance to churn in mobile telecom networks. In *Proceedings of the 11th international conference on Extending database technology: Advances in database technology* (pp. 668-677). ACM. doi:10.1145/1353343.1353424

Domingos, P. (1999). The role of Occams razor in knowledge discovery. *Data Mining and Knowledge Discovery, 3*(4), 409–425. doi:10.1023/A:1009868929893

Eiben, A. E., Koudijs, A. E., & Slisser, F. (1998). Genetic modelling of customer retention. In *Genetic Programming* (pp. 178–186). Springer Berlin Heidelberg. doi:10.1007/BFb0055937

Entner, R. (2011, June 23). *International comparisons: the handset replacement cycle*. Retrieved from http://mobilefuture.org/wp-content/uploads/2013/02/mobile-future.publications.handset-replacement-cycle.pdf

Fibich, G., & Gibori, R. I. (2010). Aggregate diffusion dynamics in agent-based models with a spatial structure. *Operations Research, 58*(5), 1450–1468. doi:10.1287/opre.1100.0818

Fisher, J. C., & Pry, R. H. (1972). A simple substitution model of technological change. *Technological Forecasting and Social Change, 3*, 75–88. doi:10.1016/S0040-1625(71)80005-7

Fourt, L. A., & Woodlock, J. W. (1960). Early prediction of market success for new grocery products. *Journal of Marketing*, *25*(2), 31–38. doi:10.2307/1248608

Goldenberg, J., Libai, B., & Muller, E. (2001). Using complex systems analysis to advance marketing theory development: Modeling heterogeneity effects on new product growth through stochastic cellular automata. *Academy of Marketing Science Review*, *9*(3), 1–18.

Hohnisch, M., Pittnauer, S., & Stauffer, D. (2008). A percolation-based model explaining delayed takeoff in new-product diffusion. *Industrial and Corporate Change*, *17*(5), 1001–1017. doi:10.1093/icc/dtn031

Hong Kong Government. (2015). *Hong Kong: The Facts*. Retrieved from http://www.gov.hk/en/about/abouthk/factsheets/docs/telecommunications.pdf

Hosmer, D. W. Jr, & Lemeshow, S. (2004). *Applied logistic regression*. John Wiley & Sons.

Hu, M., Hsieh, C., & Jia, J. (2014). *The effectiveness of peer influence and network structure: an application using mobile data*. Working Paper.

Hu, M., Yang, S., & Xu, Y. (2015). *Social Learning and Network Effects in Contagious Switching Behavior*. Working Paper.

Hung, S. Y., Yen, D. C., & Wang, H. Y. (2006). Applying data mining to telecom churn management. *Expert Systems with Applications*, *31*(3), 515–524. doi:10.1016/j.eswa.2005.09.080

International Data Corporation. (2015). *Smartphone Vendor Market Share, Q12015*. Retrieved from http://www.idc.com/prodserv/smartphone-market-share.jsp

International Telecommunication Union. (2015). *World 2015*. Retrieved from http://www.itu.int/en/ITU-D/Statistics/Documents/facts/ICTFactsFigures2015.pdf

Iyengar, R., Van den Bulte, C., & Lee, J. Y. (2015). Social contagion in new product trial and repeat. *Marketing Science*, *34*(3), 408–429. doi:10.1287/mksc.2014.0888

Iyengar, R., Van den Bulte, C., & Valente, T. W. (2011). Opinion leadership and social contagion in new product diffusion. *Marketing Science*, *30*(2), 195–212. doi:10.1287/mksc.1100.0566

Jain, A. K., Murty, M. N., & Flynn, P. J. (1999). Data clustering: A review. *ACM Computing Surveys*, *31*(3), 264–323. doi:10.1145/331499.331504

Kass, G. V. (1980). An exploratory technique for investigating large quantities of categorical data. *Applied Statistics*, *29*(2), 119–127. doi:10.2307/2986296

Katz, M. L., & Shapiro, C. (1985). Network externalities, competition, and compatibility. *The American Economic Review*, 424–440.

Kiesling, E., Günther, M., Stummer, C., & Wakolbinger, L. M. (2012). Agent-based simulation of innovation diffusion: A review. *Central European Journal of Operations Research*, *20*(2), 183–230. doi:10.1007/s10100-011-0210-y

Kisioglu, P., & Topcu, Y. I. (2011). Applying Bayesian Belief Network approach to customer churn analysis: A case study on the telecom industry of Turkey. *Expert Systems with Applications*, *38*(6), 7151–7157. doi:10.1016/j.eswa.2010.12.045

Kocsis, G., & Kun, F. (2008). The effect of network topologies on the spreading of technological developments. *Journal of Statistical Mechanics*, *2008*(10), P10014. doi:10.1088/1742-5468/2008/10/P10014

Kohonen, T. (1987). Adaptive, associative, and self-organizing functions in neural computing. *Applied Optics*, *26*(23), 4910–4918. doi:10.1364/AO.26.004910 PMID:20523469

Langley, P., Iba, W., & Thompson, K. (1992, July). *An analysis of Bayesian classifiers* (Vol. 90). AAAI.

Lazarov, V., & Capota, M. (2007). *Churn prediction*. Bus. Anal. Course. TUM Comput. Sci.

Lemmens, A., & Croux, C. (2006). Bagging and boosting classification trees to predict churn. *JMR, Journal of Marketing Research*, *43*(2), 276–286. doi:10.1509/jmkr.43.2.276

Mahajan, V., & Muller, E. (1979). Innovation diffusion and new product growth models in marketing. *Journal of Marketing*, *43*(4), 55–68. doi:10.2307/1250271

Mahajan, V., Muller, E., & Bass, F. M. (1990). New product diffusion models in marketing: A review and directions for research. *Journal of Marketing*, *54*(1), 1–26. doi:10.2307/1252170

Manchanda, P., Xie, Y., & Youn, N. (2008). The role of targeted communication and contagion in product adoption. *Marketing Science*, *27*(6), 961–976. doi:10.1287/mksc.1070.0354

Mansfield, E. (1961). Technical change and the rate of imitation. *Econometrica*, *29*(4), 741–766. doi:10.2307/1911817

Masand, B., Datta, P., Mani, D. R., & Li, B. (1999). CHAMP: A prototype for automated cellular churn prediction. *Data Mining and Knowledge Discovery, 3*(2), 219–225. doi:10.1023/A:1009873905876

Moretti, E. (2011). Social learning and peer effects in consumption: Evidence from movie sales. *The Review of Economic Studies, 78*(1), 356–393. doi:10.1093/restud/rdq014

Nambiar, R., & Poess, M. (2011). Transaction performance vs. Moore's law: a trend analysis. In Performance Evaluation, Measurement and Characterization of Complex Systems (pp. 110-120). Springer Berlin Heidelberg.

Narayan, V., Rao, V. R., & Saunders, C. (2011). How peer influence affects attribute preferences: A Bayesian updating mechanism. *Marketing Science, 30*(2), 368–384. doi:10.1287/mksc.1100.0618

Nath, S. V., & Behara, R. S. (2003, November). Customer churn analysis in the wireless industry: A data mining approach. *Proceedings-annual meeting of the decision sciences institute*, 505-510.

Neslin, S. A., Gupta, S., Kamakura, W., Lu, J., & Mason, C. H. (2006). Defection detection: Measuring and understanding the predictive accuracy of customer churn models. *JMR, Journal of Marketing Research, 43*(2), 204–211. doi:10.1509/jmkr.43.2.204

Ngai, E. W., Xiu, L., & Chau, D. C. (2009). Application of data mining techniques in customer relationship management: A literature review and classification. *Expert Systems with Applications, 36*(2), 2592–2602. doi:10.1016/j.eswa.2008.02.021

Office of the Communications Authority. (2015). Table 3: Telecommunications Services. In *Key Communications Statistics*. Retrieved from http://www.ofca.gov.hk/en/media_focus/data_statistics/key_stat/

Peres, R., Muller, E., & Mahajan, V. (2010). Innovation diffusion and new product growth models: A critical review and research directions. *International Journal of Research in Marketing, 27*(2), 91–106. doi:10.1016/j.ijresmar.2009.12.012

Quinlan, J. R. (2014). *C4. 5: programs for machine learning*. Elsevier.

Rahmandad, H., & Sterman, J. (2008). Heterogeneity and network structure in the dynamics of diffusion: Comparing agent-based and differential equation models. *Management Science, 54*(5), 998–1014. doi:10.1287/mnsc.1070.0787

Rogers, E. M. (1962). *The Diffusion of Innovations*. New York, NY: Free Press.

Rogers, E. M. (2010). *Diffusion of innovations*. Simon and Schuster.

Saylor, J. (2004, Oct 11). *The Missing Link in CRM: Customer Acquisition Management*. Retrieved from http://www.destinationcrm.com/Articles/Web-Exclusives/Viewpoints/The-Missing-Link-in-CRM-Customer-Acquisition-Management-44024.aspx

Schmittlein, D. C., & Mahajan, V. (1982). Maximum likelihood estimation for an innovation diffusion model of new product acceptance. *Marketing Science*, *1*(1), 57–78. doi:10.1287/mksc.1.1.57

Snijders, T., Steglich, C., & Schweinberger, M. (2007). Modeling the coevolution of networks and behavior. Academic Press.

Snijders, T. A. (1996). Stochastic actor - oriented models for network change. *The Journal of Mathematical Sociology*, *21*(1-2), 149–172. doi:10.1080/0022250X.1996.9990178

Snijders, T. A. (2001). The statistical evaluation of social network dynamics. *Sociological Methodology*, *31*(1), 361–395. doi:10.1111/0081-1750.00099

Srinivasan, V., & Mason, C. H. (1986). Technical note-nonlinear least squares estimation of new product diffusion models. *Marketing Science*, *5*(2), 169–178. doi:10.1287/mksc.5.2.169

Tsai, C. F., & Lu, Y. H. (2009). Customer churn prediction by hybrid neural networks. *Expert Systems with Applications*, *36*(10), 12547–12553. doi:10.1016/j.eswa.2009.05.032

Ultsch, A. (2002). Emergent self-organising feature maps used for prediction and prevention of churn in mobile phone markets. *Journal of Targeting, Measurement and Analysis for Marketing*, *10*(4), 314–324. doi:10.1057/palgrave.jt.5740056

Van den Bulte, C., & Lilien, G. L. (2001). Medical innovation revisited: Social contagion versus marketing effort. *American Journal of Sociology*, *106*(5), 1409–1435. doi:10.1086/320819

Van den Poel, D., & Lariviere, B. (2004). Customer attrition analysis for financial services using proportional hazard models. *European Journal of Operational Research*, *157*(1), 196–217. doi:10.1016/S0377-2217(03)00069-9

Von Neumann, J., & Burks, A. W. (1966). Theory of self-reproducing automata. *IEEE Transactions on Neural Networks*, *5*(1), 3–14.

Wei, C. P., & Chiu, I. T. (2002). Turning telecommunications call details to churn prediction: A data mining approach. *Expert Systems with Applications*, *23*(2), 103–112. doi:10.1016/S0957-4174(02)00030-1

## KEY TERMS AND DEFINITIONS

**Churn:** In mobile industry, churn means to discontinue the contract with the current operator.

**Customer Acquisition:** To obtain new customers.

**Customer Retention:** To keep existing customers from switching to other companies.

**Network Structure:** Topological properties of the network.

**Product Diffusion:** The process of a new product accepted by the market.

**Social Network:** The social structure in which individuals are connected with each other according to their communication patterns or other social rules.

**Targeting:** In marketing context, targeting means to choose a particular group of customers based on their characteristics.

# Chapter 8

# Anomaly Detection in Wireless Networks:
## An Introduction to Multi– Cluster Technique

**Yirui Hu**
*Rutgers University, USA*

## ABSTRACT

*This chapter is an introduction to multi-cluster based anomaly detection analysis. Various anomalies present different behaviors in wireless networks. Not all anomalies are known to networks. Unsupervised algorithms are desirable to automatically characterize the nature of traffic behavior and detect anomalies from normal behaviors. Essentially all anomaly detection systems first learn a model of the normal patterns in training data set, and then determine the anomaly score of a given testing data point based on the deviations from the learned patterns. The initial step of learning a good model is the most crucial part in anomaly detection. Multi-cluster based analysis are valuable because they can obtain the insights of human behaviors and learn similar patterns in temporal traffic data. The anomaly threshold can be determined by quantitative analysis based on the trained model. A novel quantitative "Donut" algorithm of anomaly detection on the basis of model log-likelihood is proposed in this chapter.*

DOI: 10.4018/978-1-5225-1750-4.ch008

# INTRODUCTION

An anomaly in wireless networks represents an unusual traffic pattern that deviates from the normal (or usual) network behavior. Anomalies are also referred to be abnormalities, deviants, or outliers in data mining and statistics field. Network anomalies can be root caused by a variety of issues such as introduction of new features, network intrusions or disaster events. In many cases, intrusion for example, the outliers can only be discovered as a sequence of multiple data points, rather than as an individual data point.

The general steps in developing an anomaly detection system are: first define a model of the normal patterns using training data; then compute the deviation score of testing data points given normal patterns (Aggarwal, 2013). Training a model for co-occurrence data includes both manually and automatic techniques. In past few decades, the network alarm thresholds can be set manually and tuned by subject matter experts (SME) for each generating entity. In nowadays network traffic, however, it is almost impossible for SMEs to manually detect anomalies from such voluminous amount of high-dimensional data. Effective anomaly detection systems based on machine learning algorithms are hence desirable to automatically extract useful information in terms of abnormal characteristics of the systems and entities from such noisy, high-dimensional data and provide useful application-specific insights.

It is therefore crucial to learn a good model based on training data. The optimal choice of a model is often data set specific, which requires a good understanding of the particular domain before choosing the model (Aggarwal, 2013). Clustering analysis can be a promising candidate in learning similar patterns in temporal traffic data.

# BACKGROUND

The initial step of learning a good model is the most crucial part in anomaly detection. In general, an incorrect choice of model can lead to poor anomaly detection results. For example, a linear model may not work well if the underlying pattern is generated from multiple clusters. In such cases, the testing data can be mistakenly detected as anomaly because the poor fit to the learned linear model, which lead to high false alarm rates. Effective anomaly detection systems based on machine learning algorithms are hence desirable to automatically extract useful information in terms of abnormal characteristics of the systems and entities from such noisy, high-dimensional data and provide useful application-specific insights.

Machine learning algorithms include supervised learning and unsupervised learning (Jain, Murty, & Flynn, 1999; Theodoridis & Koutroumbas, 2006) . Most anomaly detection systems employ supervised algorithms based on training data, however, the training data are typically expensive to generate (Leung & Leckie, 2005). Moreover, these supervised detection techniques have difficulty in detecting new types of anomaly.

In the absence of unlabeled data, unsupervised anomaly detection techniques can be promising because they are capable of detecting previously unseen anomalies. From unsupervised learning, we are able to learn particular patterns in a way that reflects the statistical structure of the overall system. Unsupervised learning has been widely used in history. Hebb (1949) linked statistical methods to experiments. Hinton and Sejnowski (1986) invented the Boltzmann machine model, which provided insights in the density estimation methods. Some early work include Horace Barlow (1992), Donald MacKay (1956), and David Marr (1970).

Most anomaly detection systems derive network traffic model from training stage. In Brutlag (2000), Brutlag uses an extension of Holt Winter algorithm to capture the history of the network traffic variations and to predict for future traffic rate using confidence band. The incremental model updates via exponential smoothing. When the captured feature of network traffic continues to fall outside of the confidence band, an alarm is raised. In Barfiord, Kline, Plonka, and Shim (2000), Barfiord et al. implement wavelet analysis to remove the predictable ambient part from the traffic and then study the variation part in network traffic rate. Network anomalies are therefore detected by applying a threshold to a deviation score based on the trained model with normal patterns.

From another perspective, clustering has attracted interest from researchers in finding abnormal patterns (Ramaswamy, Rastogi, & Shim, 2000; Sequeira & Zaki, 2002). The main advantage of clustering based algorithms is the ability to learn normal patterns and detect anomalies in the unlabeled data, while not requiring explicit descriptions of various anomaly types (Patcha & Park, 2007). Modern techniques include hard clustering and soft clustering. In hard clustering each data point belongs exactly to one group, while in soft clustering a data point can belong to more than one group. The most popular and simplest clustering technique is *K*-means hard clustering (Jain, 2010). The idea is to partition n observations into *K* clusters in which each observation belongs to the cluster with the nearest mean. A significant disadvantage of *K*-Means is the *K* must be pre-defined with domain knowledge. A sad fact is *K* has no way to be confirmed before training in many cases. The main advantage of soft clustering techniques over hard cluster lies in the assumption that same observation can belong to different clusters, which thus allows for more flexibility in probabilistic modeling. Advanced methods of parametric statistical detection including probabilistic/GMM approaches, which is a well-known soft clustering technique that assumes all the data points are generated from a mixture of finite number Gaussian distribution. In Haiji (2005), Hajji applies a Gaussian Mixture model (GMM) and develops a stochastic approximation of the Expectation-Maximization (EM) algorithm to estimate parameters.

GMMs are commonly used as a parametric model of the probability distribution of continuous measurements or features in a biometric system, such as vocal-tract related spectral features in a speaker recognition system (Reynolds, 1992). GMMs are most notably in speaker recognition systems, due to their capability of representing a large class of sample distributions. One of the powerful attributes of the GMM is its ability to form smooth approximations to arbitrarily shaped densities (McLachlan, 1988). Each component Gaussian density represents feature distributions by a position (mean vector) and an elliptic shape (covariance matrix).

Note that A GMM can also be viewed as a single-state Hidden Markov Model (HMM) with a Gaussian mixture observation density, or an ergodic Gaussian observation HMM with fixed, equal transition probabilities. Assuming independent feature vectors, the observation density of feature vectors drawn from these hidden acoustic classes is a Gaussian mixture (Reynolda, 1992; McLachlan, 1988).

## MAIN FOCUS OF THE CHAPTER

The optimal choice of model usually comes with a good understanding of domain knowledge. Clustering models are of value because they have the ability to identify the underlying similar patterns in temporal traffic data. In specific, multiple cluster-based analytical models are recommended in such cases as they can obtain insights into the mathematical structure of each cluster (that incorporates similar distributed generating entities) and the relationships to an individual data point.

The use of a GMM for representing feature distributions in a network may also be motivated by the intuition that the individual component densities may model some underlying set of hidden classes. For example, in cellular traffic data, it is reasonable to assume the temporal patterns of related features corresponding to classes of user behavior activities. When people wake up in the morning, the morning value for features begin to increase; As people go to work, the daytime value for features reach the peak; When people go to bed, the evening value decrease until reach the bottom. Each component of Gaussian densities indicates the underlying set of hidden class that characterizes the user behaviors pattern. The shape of the $i^{th}$ class can in turn be represented by the mean vector ($\mu_i$), and variations by the covariance matrix ($\Sigma_i$) of each component Gaussian.

## ISSUES, CONTROVERSIES, PROBLEMS

As we discussed earlier, traffic data has the temporal patterns at the macro scale. Mobile activity rapidly grows in the morning between 6 and 10 AM, followed by rather steady activity levels during a day, and a slower decrease at night between 9 PM and 2 AM (Grauwin, Sobolevsky, Moritz, Godor, & Ratti, 2015). It is reasonable to associate those temporal patterns to a daily cycle, corresponding to people waking up in the morning, working during the daytime, and then going home in the evening.

One can develop specific model for every discrete time slot. Common assumptions could be univariate or multivariate Gaussian distribution (UVG, MVG) for sub-dataset of each time slot. Therefore, if observations are measured half an hour, then we have 48 UVG models in total. We can conduct anomaly detection using Mahalanobis distance as in Eskin, Amold, Prerau, Portnoy, and Stolfo (2002). This type of models is easy to conduct, however, it has the following shortcomings: i) The parameters of the distribution learned from the segmentation of data is of poor quality since there is much less data per segment to train with. ii) It is much less efficient as it does not consider the fact that several time slots may have similar behavior and may be pooled together. iii) It ignores the time correlation between adjacent time series points.

The challenge of traffic data analysis is to identify the underlying similar patterns in temporal data without modeling each time slot independently.

## SOLUTIONS AND RECOMMENDATIONS

As we stated earlier, the initial step of learning a good model is the most crucial part in anomaly detection. Multiple cluster-based analytical models are recommended in such cases as they can obtain insights into the mathematical structure of each cluster (that incorporates similar distributed generating entities) and the relationships to an individual data point.

In this part, we first divide the data into training and testing data and obtain trained GMM parameters from training data, then compute the likelihood score of testing data given the trained parameters for anomaly detection.

A Gaussian Mixture Model is a probabilistic model with a weighted sum of $K$ component Gaussian densities as defined below.

$$p\left(x \mid \lambda\right) = \sum_{i=1}^{K} \omega_i g(x \mid \mu_i, \Sigma_i),$$ (1)

where $x$ is a $D$-dimensional continuous-valued data vector (i.e. features), $\omega_i$, $i=1$, ..., $K$, are the mixture weights, and $g(x \mid \mu_i, \Sigma_i)$, $i=1$, ..., $K$, are the component Gaussian densities. Each component density is a $D$-variate Gaussian function of the the form,

$$g\left(x \mid \mu_i, \Sigma_i\right) = \frac{1}{\left(2\pi\right)^{D/2} \left|\Sigma_i\right|^{1/2}} \exp\left\{-\frac{1}{2}\left(x - \mu_i\right)' \Sigma_i^{-1}\left(x - \mu_i\right)\right\}, \tag{2}$$

with $D$-dimensional mean vector $\mu_i$ and $D$ by $D$ covariance matrix $\Sigma_i$. The mixture weights satisfy the constraint that $\sum_{i=1}^{K}\omega_i = 1$. The parameter set for a complete GMM includes the mean vector, covariance matrix and the mixture weights.

$$\lambda = \left\{\omega_i, \mu_i, \Sigma_i\right\}, i=1, ..., K \tag{3}$$

The second step of anomaly detection is to determine the anomaly score of a given testing data point based on the deviations from the learned patterns. According to the definition from Hawkins (1992), an anomaly is an observation which deviates too much from the others, and it could be possibly generated by a different mechanism. For numerical data, outliers are defined as data points which do not belong to any members of clusters, or are not in sufficient proximity of clusters.

The most basic form of anomalies is extreme value analysis (Ahmed, Oreshkin, & Coates, 2007). The key for extreme value analysis is to determine the statistical tails for given trained model. Any observation goes beyond the statistical tails are considered as anomalies. Although it is natural to conduct extreme value analysis to univariate data, it may not be applied directly to multivariate data. One possible solution is to estimate the weighted average multiple anomaly scores into one single score. However, it may not easy to determine the weighted score for multivariate data.

This chapter introduces a more general quantitative approach using model likelihood. Model likelihood is defined as the probability of observed data given a model with parameters. Observation with a relatively low likelihood is more likely to be an anomaly.

Given training data and the GMM configuration, we can estimate the parameter set $\lambda = \left\{\omega_i, \mu_i, \Sigma_i\right\}$, $i=1$, ..., $K$ using Maximum Likelihood Estimation (MLE), which best matches the distribution of the training feature vectors. The aim of MLE is to find the model parameters that maximize the likelihood of GMM given the training data. For a sequence of N training vectors $X = \left\{x_1, ..., x_N\right\}$, the GMM likelihood assuming independency among training vectors $x_l, l = 1, \ldots, N$, can be written as,

$$p\left(X \mid \lambda\right) = \prod_{l=1}^{N} p\left(x_l \mid \lambda\right). \tag{4}$$

MLE can be obtained iteratively using Expectation Maximization (EM) algorithm (Dempster, Laird, & Rubin, 1977). On each EM iteration, the following re-estimation formulas are used to guarantee a monotonically increasing model likelihood value.

$$\hat{\omega}_i = \frac{1}{N} \sum_{l=1}^{N} \Pr(i \mid x_l, \lambda), \tag{5}$$

$$\hat{\mu}_i = \frac{\sum_{l=1}^{N} \Pr(i \mid x_l, \lambda) x_l}{\sum_{l=1}^{N} \Pr(i \mid x_l, \lambda)}, \tag{6}$$

$$\hat{\sigma}_i^2 = \frac{\sum_{l=1}^{N} \Pr(i \mid x_l, \lambda) x_l^2}{\sum_{l=1}^{N} \Pr(i \mid x_l, \lambda)} - \hat{\mu}_i^2. \tag{7}$$

Having learned the GMM parameter set, we are able to detect anomaly in testing data. There are a bunch of ways to define anomaly using different statistics and tests. In this paper, we introduce a novel "Donut" algorithm for detect abnormal patterns in test data.

The idea is to discover the unusual likelihood of test data given trained parameters. Applying trained parameters to test data yields the likelihood of the occurrence. The steps are as follows:

1.  Estimate parameters of above models using EM algorithm in train data.
2.  Calculate log-likelihood of train data, and find 10% and 1% quantiles as thresholds.
3.  Apply trained model using online test data, and divide their log-likelihoods into 3 zones:

    a.    **Red Zone:** Any region below the 1% quantile is considered as red zone, which is dangerous since it contains extreme unlikely values.

    b.    **Green Zone:** Any region above the 10% quantile is considered as green zone, which is safe since it's very likely given the trained model.

    c.    **Yellow Zone:** Any region in between is defined as yellow zone, which is composed of unlikely values and requires further attention. If the log-likelihood stay in yellow region (or even red region) for a certain window, say $w=6$, then the starting test data point is labeled as anomaly.

4.    Determine the types of anomaly for test data using 3 zones:

    a.    If the log-likelihood falls into the red zone, then label the test data as Type 1 anomaly.

    b.    If the testing log-likelihood fell into the yellow zone and persisted for a certain period (window length=6), then label the starting data as Type 2 anomaly.

The advantage of "Donut" algorithm is that it can automatically detect two types of anomaly: Type 1 anomalies are single traffic data points with extreme unlikely log-likelihood given trained model; Type 2 anomalies are unlikely traffic data series whose log-likelihoods persist within the unlikely thresholds in a certain length of window.

## FUTURE RESEARCH DIRECTIONS

There is a trade-off between the accuracy and complexity of a model. One of the fundamental problems in clustering problem is how to choose the number of Gaussian components ($K$). Analytical models with smaller values of $K$ are easy to compute, however, they may not explicitly model the data; models with a larger $K$ can better learn and explain the data with more clusters, however, the EM algorithms are harder to converge as the number of parameters increases dramatically. Also an over-fitting problem may occur as $K$ increases. In summary, a general model with too many parameters may have the problem of over-fitting and therefore be sensitive to anomalies; a simple model may poorly fit the data and thus mistakenly declare normal data as anomalies. The Gaussian Mixture Model is robust when $K$ is relative large, which is able to identify anomalies with a low false alarm rate.

Future network analysis is to develop a self-diagnosing system that triggers self-optimizing and self-healing wherein anomaly detection plays a critical role. It is hence clear that anomaly detection must be accurate and sensitive.

# CONCLUSION

Various anomalies present different behaviors in wireless networks. Not all anomalies are known to networks. Essentially all anomaly detection systems first learn a model of the normal patterns in training data set, and then determine the anomaly score of a given testing data point based on the deviations from the learned patterns. The initial step of learning a good model is the most crucial part in anomaly detection. Several clustering-based unsupervised machine learning algorithms for traffic data are discussed in this chapter. GMMs are recommended in temporal traffic data as they can obtain insights into the mathematical structure of each cluster (that incorporates similar distributed generating entities) and the relationships to an individual data point. The GMM is robust when $K$ is relative large, which is able to identify anomalies with a low false alarm rate.

# REFERENCES

Aggarwal, C. C. (2013). *Outlier Analysis*. New York: Springer. doi:10.1007/978-1-4614-6396-2

Ahmed, T., Oreshkin, B., & Coates, M. J. (2007). April). Machine learning approaches to network anomaly detection. In *Proceedings of the 2nd USENIX workshop on Tackling computer systems problems with machine learning techniques* (pp. 1-6). USENIX Association.

Barfiord, P., Kline, J., Plonka, D., & Ron, A. (2002). A signal analysis of network traffic anomalies. In *Proceedings of ACM SIGCOMM Internet Measurement Workshop*. New York: ACM. doi:10.1145/637201.637210

Barlow, H. B. (1989). Unsupervised learning. *Neural Computation*, *1*(3), 295–311. doi:10.1162/neco.1989.1.3.295

Brutlag, J. D. (2000). Aberrant behavior detection in time series for network service monitoring. In *Proceeding of the 14th Systems Administration Conference* (pp. 139-146). USENIX Association.

Dempster, A., Laird, N., & Rubin, D. (1977). Maximum Likelihood from Incomplete Data via the EM Algorithm. *Journal of the Royal Statistical Society. Series A (General)*, *39*(1), 1–38.

Eskin, E., Arnold, A., Prerau, M., Portnoy, L., & Stolfo, S. (2002). A geometric framework for unsupervised anomaly detection. In Applications of data mining in computer security (pp. 77-101). Springer US. doi:10.1007/978-1-4615-0953-0_4

Grauwin, S., Sobolevsky, S., Moritz, S., Gódor, I., & Ratti, C. (2015). Towards a comparative science of cities: Using mobile traffic records in new york, london, and hong kong. In *Computational approaches for urban environments* (pp. 363–387). Springer International Publishing. doi:10.1007/978-3-319-11469-9_15

Hajji, H. (2005). Statistical analysis of network traffic for adaptive faults detection. *IEEE Transactions on Neural Networks*, *16*(5), 1053–1063. doi:10.1109/TNN.2005.853414 PMID:16252816

Hebb, D. (1949). *The Organization of Behavior*. New York: Wiley.

Hinton, G. E., & Sejnowski, T. J. (1986). Learning and relearning in Boltzmann machines. In Parallel Distributed Processing: Explorations in the Microstructure of Cognition:*Foundations* (vol. 1, pp. 282–317). Cambridge, MA: MIT Press.

Jain, A., Murty, M., & Flynn, P. (1999). Data clustering: A review. *ACM Computing Surveys*, *31*(3), 264–323. doi:10.1145/331499.331504

Jain, A. K. (2010). Data clustering: 50 years beyond K-means. *Pattern Recognition Letters*, *31*(8), 651–666. doi:10.1016/j.patrec.2009.09.011

Leung, K., & Leckie, C. (2005). Unsupervised anomaly detection in network intrusion detection using clusters. In *Proceedings of the Twenty-eighth Australasian conference on Computer Science* (Vol. 38, pp. 333-342). Australian Computer Society, Inc.

MacKay, D. M. (1956). The epistemological problem for automata. In C. E. Shannon & J. McCarthy (Eds.), *Automata Studies* (pp. 235–251). Princeton, NJ: Princeton University Press.

Marr, D. (1970). A theory for cerebral neocortex. *Proceedings of the Royal Society of London. Series B, Biological Sciences*, *176*(1043), 161–234. doi:10.1098/rspb.1970.0040 PMID:4394740

McLachlan, G. J., & Basford, K. E. (1988). *Mixture models. Inference and applications to clustering*. New York: Dekker.

Patcha, A., & Park, J. M. (2007). An overview of anomaly detection techniques: Existing solutions and latest technological trends. *Computer Networks*, *51*(2), 3448–3470. doi:10.1016/j.comnet.2007.02.001

Ramaswamy, S., Rastogi, R., & Shim, K. (2000). Efficient algorithms for mining outliers from large data sets. In *ACM SIGMOD international conference on Management of data* (pp. 427–438). New York: ACM. doi:10.1145/342009.335437

Reynolds, D. A. (1992). *A Gaussian Mixture Modeling Approach to Text-Independent Speaker Identification* (PhD thesis). Georgia Institute of Technology.

Sequeira, K., & Zaki, M. (2002). ADMIT: Anomaly-based data mining for intrusions. In *8th ACM SIGKDD international conference on Knowledge discovery and data mining* (pp. 386-395). New York: ACM.

Theodoridis, S., & Koutroumbas, K. (n.d.). *Pattern recognition* (3rd ed.). Academic Press.

## KEY TERMS AND DEFINITIONS

**Anomaly Detection:** The identification of items, events or observations which do not conform to an expected pattern or other items in a dataset.

**Cluster Analysis:** The task of grouping a set of objects in such a way that objects in the same group (called a cluster) are more similar to each other than to those in other groups (clusters).

**Model Log-Likelihood:** A likelihood is a function of the parameters of the statistical model given data, and a log-likelihood is the natural logarithm of the likelihood function.

**Unsupervised Learning:** A type of machine learning algorithm used to draw inferences from datasets consisting of input data without labeled responses.

# Chapter 9

# Continuous–Time Markov Chain–Based Reliability Analysis for Future Cellular Networks

**Hasan Farooq**
*University of Oklahoma, USA*

**Md Salik Parwez**
*University of Oklahoma, USA*

**Ali Imran**
*University of Oklahoma, USA*

## ABSTRACT

*It is anticipated that the future cellular networks will consist of an ultra-dense deployment of complex heterogeneous Base Stations (BSs). Consequently, Self-Organizing Networks (SON) features are considered to be inevitable for efficient and reliable management of such a complex network. Given their unfathomable complexity, cellular networks are inherently prone to partial or complete cell outages due to hardware and/or software failures and parameter misconfiguration caused by human error, multivendor incompatibility or operational drift. Forthcoming cellular networks, vis-a-vis 5G are susceptible to even higher cell outage rates due to their higher parametric complexity and also due to potential conflicts among multiple SON functions. These realities pose a major challenge for reliable operation of future ultra-dense cellular networks in cost effective manner. In this paper, we present a stochastic analytical model to analyze the effects of arrival of faults in a cellular*

DOI: 10.4018/978-1-5225-1750-4.ch009

*network. We exploit Continuous Time Markov Chain (CTMC) with exponential distribution for failures and recovery times to model the reliability behavior of a BS. We leverage the developed model and subsequent analysis to propose an adaptive fault predictive framework. The proposed fault prediction framework can adapt the CTMC model by dynamically learning from past database of failures, and hence can reduce network recovery time thereby improving its reliability. Numerical results from three case studies, representing different types of network, are evaluated to demonstrate the applicability of the proposed analytical model.*

## I. INTRODUCTION

The envisioned fifth Generation (5G) Cellular Networks are expected to achieve 1000 times capacity gain mainly through extreme network densification (Figure 1) (Imran, Zoha & Abu-Dayya, 2014). Moreover, with each successive generation of cellular networks, complexity of BS has continued to increase i.e. a typical 2G cell has 500 parameters to optimally configure and maintain; 3G cell has 1000; and 4G has roughly 1500 parameters. Without intervening measures, same complexity growth trend is expected for 5G (Imran et al., 2014). To efficiently manage such an ultra-dense, complex, heterogeneous cellular networks, the paradigm of SON has recently been

*Figure 1. Ultra dense heterogeneous complex cellular network*

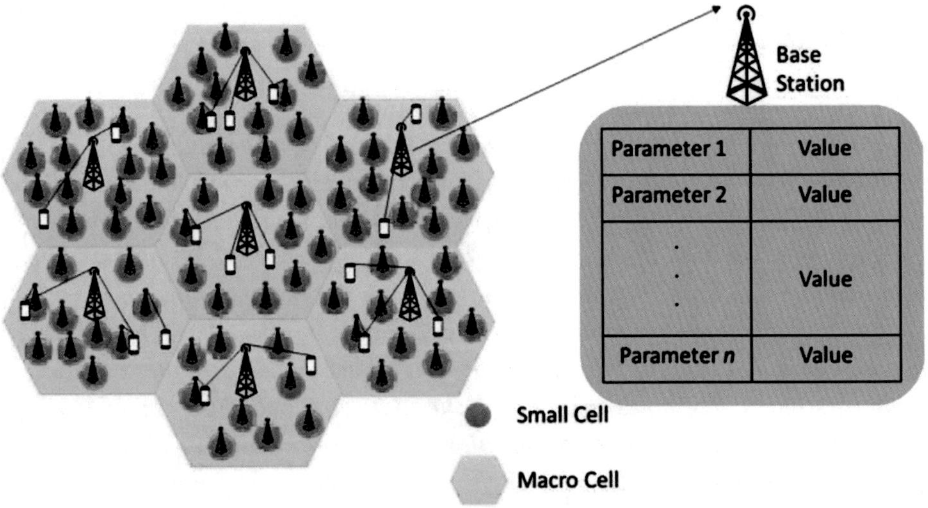

investigated heavily to automate network configuration and management tasks (Hämäläinen, Sanneck & Sartori, 2012). Realizing the importance of SON as a key enabler for future cellular networks, a number of SON use cases have already been standardized by 3GPP (European Telecommunications Standards Institute, 2014). There is a general consensus that SON will not be a luxury but a necessity in 5G networks (Østerbø & Grøndalen, 2012). SON functions, classified as

1.  Self-Configuring,
2.  Self-Optimizing, and
3.  Self-Healing.

They operate by reconfiguring a number of network parameters. For a thorough review of state of the art SON function see Aliu, Imran, Imran & Evans (2013).

Cellular networks are inherently subject to cell outages caused by either BS hardware and/or software malfunctions or misconfiguration of several hundred cell parameters during routine network operation. Forthcoming cellular networks are susceptible to even higher cell outages rates as the multiple SON functions may be subjected to large number of potential conflicts when operated concurrently in a system. Given the parametric overlap as well as coupling among the objectives of different SON functions, it has been shown in Lateef, Imran & Abu-Dayya (2013) that large number of conflicts are possible among SON functions if no self-coordination mechanism is employed. At times, such conflicts can actually degrade networks performance instead of improving it. For example capacity and coverage optimization SON function might try to improve coverage by increasing transmission power. This may conflict with energy efficiency SON. In summary, as identified in Lateef et al. (2013) in an uncoordinated SON, a variety of conflicts may occur when:

1.  Two or more SON functions try to modify the same network configuration parameter.
2.  A SON function is triggered by an input parameter whose value is dependent upon some other network parameters.
3.  There is a change in network conditions by impromptu addition or removal of relay, eNB or Home eNB (HeNB).
4.  Different SON function actions try to alter the same KPI of a cell, while adjusting different network configuration parameters 5) A SON function computes new parameter configuration values based on outdated measurements.
5.  There is a logical dependency among the objectives of SON functions.

The potential failures occurring due to hardware/software malfunctioning, multi-vendor incompatibility or SON conflicts ultimately affect the coverage and performance reliability of the network. BS can be susceptible to complete outage due to critical failures or can exhibit degraded performance in case of trivial failures. The reliability analysis of future cellular network's BS is of paramount importance for network operators since it directly effects the Quality of Service and user experience. A quantitative analysis of SON reliability can also give vendors a better insight into the various reliability considerations in SON. It can also help improve operator's confidence on SON which has been major bottleneck in SON penetration despite of the enormous financial and technical gain SON can offer. Despite the great significance of the topic, so far very few studies have focused on the reliability and survivability analysis of cellular networks in general and SON enabled cellular networks in particular. Dharmaraja, Jindal and Varshney (2008) developed analytical model for reliability and survivability quantification of a UMTS architecture network. Xie, Heegaard and Jiang (2013) modeled and analyzed the survivability of an infrastructure based wireless network under disaster propagation. Tipper, Charnsripinyo, Shin and Dahlberg (2002) did simulation based survivability analysis of a mobile network. Guida, Longo and Postiglione (2010) evaluated performance analysis of IP Multimedia Subsystem (IMS) core network signaling servers. However, unlike the previous works on cellular network reliability that mostly focus on the structural aspects of cellular networks and overlook the network behavioral aspects that can cause complete or partial failures, our work is more focused towards developing a generic analytical model encompassing diverse faults cases like software/hardware failures or SON conflict attributed misconfigurations. This approach gives the flexibility to incorporate variety of failure scenarios into the model. To the best of our knowledge a study that analyzes the probabilistic reliability behavior of the SON enabled emerging cellular networks including 5G by considering the failure probability of BSs using CTMC does not exist in open literature. This paper is first attempt in that direction. Rest of this paper is organized as follows: Section II presents the reliability behavior model of the BS. In Section III, we have performed Transient analysis along with the computation of performance metrics. Numerical results are presented in Section IV while the utility of the developed model is presented in Section V. In Section VI, key conclusions and Future works are discussed.

## II. MODEL DEVELOPMENT

To analyze and evaluate the reliability behavior of a cellular network, a quantitative model for a cell (BS) failure is needed. In real-world cases, most of the node failure and repair times follow time-dependent failure rate distributions such as Weibull, Pareto and lognormal (Pham, 2007). However, in most cases, analytical models with general (non-exponential) distributions are not mathematically tractable. Therefore, phase-type distribution, which is convolution of many exponential phases is used for approximating many general distributions and is used to construct the mathematically solvable analytical models (Dharmaraja et al., 2008), for component reliability analysis (Osogami & Harchol-Balter, 2006). Since exponential distribution is a particular case of phase type distribution, hardware and software faults are commonly modeled as exponential distribution. Therefore in this paper, we consider that the time to transit from a system state to another due to failures and recovery also follows an exponential distribution. This assumption is also supported by the fact that the exponential random variable is the only continuous random variable with Markov property. Building on this assumption we construct an analytically tractable CTMC model for reliability analysis of SON enabled BS. Figure 2 shows the state transition diagram of the CTMC model for the probabilistic reliability behavior of the BS.

Let $X(t)$ with finite State Space $S = \{1, 2, 3\}$ denote the state of the BS at time t wherein:

- $X(t) = 1$ if BS is in healthy state at time $t$ with all parameters configured with optimal value.
- $X(t) = 2$ if BS is in sub-optimal state at time $t$ with one or more of the parameters misconfigured. In this state, cell continue to operate but its performance degrades below a typical level of performance.
- $X(t) = 3$ if BS is in complete outage at time $t$.

It is assumed that time for failure is exponentially distributed. Since the rate of arrival of failures is temporarily independent, it can be modelled using Poisson distribution. We classify failures into (1) Trivial failures characterized by arrival rate $\lambda_t$. Trivial failures are the failures which do not cause complete outage but drive the network state from optimal to Sub-optimal state. (2) Critical failures characterized by arrival rate $\lambda_c$. Critical failures lead to complete outage of the cell. Therefore, trivial failures can only occur when network is in optimal state while critical failures can occur in state 1 or state 2. Each BS is assumed to be equipped with recovery module powered by self-coordination framework such as proposed in Lateef et al. (2013). This module reconfigures all configuration parameters back to their original

*Figure 2. State transition diagram for SON enabled BS*

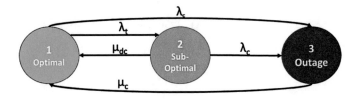

optimal values once the misconfiguration is detected and diagnosed. Moreover it has the capability to reset the BS software or switch over to the secondary backup hardware board if failure has stemmed from hardware/software related issues. The time to move the

network from sub-optimal state back to optimal state is assumed to be exponentially distributed with mean value $1/\mu_{dc}$. This includes the time for cell anomaly detection, diagnosis and compensation (Zoha et al., 2015; Wang et al., 2014). Similarly the time period to recover from complete outage is exponentially distributed with mean value $1/\mu_c$ which generally involves time for compensation only. Furthermore, the failure or repair transition is only determined by the current state and not on the path to the current state. With these assumptions, the transient process $X(t)$ can be mathematically modeled as a temporally homogeneous CTMC on the state space $S$. For each time $t > 0$, the probability that the BS is in state $j$ is given by:

$$p_j\left(t\right) = \Pr\left\{X\left(t\right) = j\right\}, j \in S \tag{1}$$

Let $p\left(t\right) = \left[p_1\left(t\right), p_2\left(t\right), p_3\left(t\right)\right]$ denote row vector of transient state probabilities of $X(t)$. The generator matrix $Q$ and rate matrix $R$ for this CTMC $X(t)$ are given as:

$$Q = \begin{bmatrix} -\lambda_t - \lambda c & \lambda_t & \lambda_c \\ \mu_{dc} & -\mu_{dc} - \lambda_c & \lambda_c \\ \mu_c & 0 & -\mu_c \end{bmatrix} \tag{2}$$

$$R = \begin{bmatrix} 0 & \lambda_t & \lambda_c \\ \mu_{dc} & 0 & \lambda_c \\ \mu_c & 0 & 0 \end{bmatrix} \tag{3}$$

## 1. Analysis

In this section we perform Transient Analysis followed by the computation of performance metrics.

## A. Transient Analysis

Using generator matrix $Q$, the dynamic behavior of the CTMC can be described by the Kolmogorov differential equation in the matrix form:

$$\frac{dP(t)}{dt} = P(t)Q \tag{4}$$

Then the transient state probability vector can be obtained as:

$$P(t) = P(0)e^{Qt} \tag{5}$$

where $P(0)$ is initial probability vector. The common methods to obtain the transient probability vector $P(t)$ includes matrix exponential approach (Trivedi, 2008) and uniformization (Kulkarni, 2011). In this paper, we resort to uniformization method for the analysis because of its higher accuracy and efficient computation due to which it is the method of choice for typical problems similar to one under consideration in this paper (Reibman & Trivedi, 1988).

Let $q_{ii}$ be the diagonal element of $Q$ and $I$ be the unit matrix, then the transient state probability vector can be obtained as follows:

$$P(t) = \sum_{k=0}^{\infty} e^{-\beta t} \frac{(\beta t)^k}{k!} \hat{P}^k \tag{6}$$

where $\beta \geq \max_i |q_{ii}|$ is uniform rate parameter and $\hat{P}$ is Probability Transition Matrix given as:

$$\hat{P} = I + Q / \beta \tag{7}$$

When we truncate the summation in (6) from infinity to some large number $M$, the resulting controllable accuracy error can be computed as:

$$\epsilon = 1 - \sum_{k=0}^{M} e^{-\beta t} \frac{(\beta t)^k}{k!} \tag{8}$$

## B. Performance Metrics

Based on the uniformization method, three performance metrics to quantify the reliability of network are computed as follows:

1.  **Occupancy Time:** The expected length of time the BS spends in each of the states {Optimal, Suboptimal, Outage} during a given interval of time can be determined using occupancy time of the CTMC. Let $\Psi_{i,j}(T)$ be the expected amount of time the CTMC spends in state $j$ during the interval $[0, T]$, starting in state $i$ and $p_{i,j}(t)$ be the element of the transition probability matrix $\hat{P}$. The quantity $\Psi_{i,j}(T)$ is called the occupancy time of state $j$. until time $T$ starting from state $i$ given as:

$$\Psi_{i,j}\left(T\right) = \int_0^T p_{i,j}\left(t\right)dt \tag{9}$$

and in matrix form:

$$\Psi\left(T\right) = \begin{bmatrix} \Psi_{1,1} & \Psi_{1,2} & \Psi_{1,3} \\ \Psi_{2,1} & \Psi_{2,2} & \Psi_{2,3} \\ \Psi_{3,1} & \Psi_{3,2} & \Psi_{3,3} \end{bmatrix} \tag{10}$$

2.  **First Passage Time:** The expected value of the random time at which BS passes into each of the states {Optimal, Suboptimal, Outage} for the first time can be calculated using first passage times of the CTMC. The first-passage time $\zeta_j$ into state $j$ starting from state $i$ is defined to be:

$$\zeta_j = E(T \mid X\left(0\right) = i) \tag{11}$$

where

$$T = \min\left\{t \geq 0 : X\left(t\right) = j\right\} \tag{12}$$

and $E$ is the expected value. The first passage times for a CTMC with a State Space $S$ satisfy the following relation (Kulkarni, 2011):

$$r_i \zeta_i = 1 + \sum_{j=1}^{N-1} r_{i,j} \zeta_j, 1 \leq i \leq N - 1 \tag{13}$$

where $i, j \in S$ and $r_i = \sum_{j=1}^{N} r_{i,j}$, $R = \left[ r_{i,j} \right]$.

Therefore in our model, first passage time for state 2 will be:

$$\left( \lambda_t + \lambda_c \right) \zeta_1 = 1 + \lambda_c \zeta_3 \tag{14}$$

$$\mu_c \zeta_3 = 1 + \mu_c \zeta_1 \tag{15}$$

By solving (14) and (15) we get:

$$\zeta_{3 \to 2} = \left( \frac{\left( \lambda_t + \lambda_c \right) + \mu_c}{\left( \left( \lambda_t + \lambda_c \right) * \mu_c \right) - \lambda_c \mu_c} \right) \tag{16}$$

$$\zeta_{1 \to 2} = \left( \frac{\mu_c \zeta_{3 \to 2} - 1}{\mu_c} \right) \tag{17}$$

where $\zeta_{3 \to 2}$ and $\zeta_{1 \to 2}$ are the first passage times to state 2 starting from states 3 and 1 respectively. First Passage time for State 3 will be:

$$\left( \lambda_t + \lambda_c \right) \zeta_1 = 1 + \lambda_t \zeta_2 \tag{18}$$

$$(\mu_{dc} + \lambda_c) \zeta_2 = 1 + \mu_{dc} \zeta_1 \tag{19}$$

By solving (18) and (19) we get:

$$\zeta_{1\to3} = \frac{\mu_{dc} + \lambda_c + \lambda_t}{(\lambda_t + \lambda_c)(\mu_{dc} + \lambda_c) - \lambda_t\mu_{dc}} \tag{20}$$

$$\zeta_{2\to3} = \frac{1 + \mu_{dc}\zeta_{1\to3}}{(\mu_{dc} + \lambda_c)} \tag{21}$$

where $\zeta_{1\to3}$ and $\zeta_{2\to3}$ are the first passage times to state 3 starting from states 1 and 2 respectively.

3.  **Steady State Distribution:** In order to analyze the long term behavior of the network, we evaluate the limiting distribution of this CTMC. The limiting or steady state distribution $\psi$ is defined as:

$$\psi = \left[\psi_1, \psi_2, \psi_3\right] \tag{22}$$

where

$$\psi_j = \lim_{t\to\infty} \Pr\left(X(t) = j\right) \tag{23}$$

For a CTMC with rate matrix $\boldsymbol{R} = [r_{i,j}]$, it is calculated as:

$$\psi_j r_j = \sum_{i=1}^{N} \psi_i r_{i,j} \tag{24}$$

and

$$\sum_{i=1}^{N} \psi_i = 1 \tag{25}$$

Therefore for our model we determine $\begin{bmatrix} \psi_1 & \psi_2 & \psi_3 \end{bmatrix}$ by solving:

$$\boldsymbol{A}\psi = \boldsymbol{B} \tag{27}$$

where

$$A = \begin{bmatrix} \lambda_t + \lambda_c & -\mu_{dc} & -\mu_c \\ \lambda_t & -(\mu_{dc} + \lambda_c) & 0 \\ \lambda_c & 0 & -\mu_c \\ 1 & 1 & 1 \end{bmatrix} \text{ and } B = \begin{bmatrix} 0 \\ 0 \\ 0 \\ 1 \end{bmatrix}$$

## II. NUMERICAL RESULTS

For numerical results, we considered three case studies with parameter settings as shown in Table 1. In Case Study I, trivial failures are assumed to occur with mean value of 8 hours in relation to the traffic pattern shifts during morning, evening and night times which might trigger a number of SON functions. Probability of occurrence of critical failures is assumed to be 10 time less than those of trivial failures. Normally cell outage detection is not a straight forward task and it may take several hours for detection, diagnosis and compensation of outages. Therefore we considered $\mu_{dc}$ to be exponentially distributed with mean value of 6 hours. In case study I, compensation is assumed to have mean value of 5 minutes and also has exponential distribution. This small recovery time makes sense only when it is assumed that the SON self-healing functions involving automated diagnosis, such as proposed in Zoha et al. (2015) and Wang et al. (2014) will be invoked for the recovery process. Otherwise, a recovery time can be significantly large. In case study II, we increased the arrival rate of misconfigurations (trivial faults) from one per eight hour to one per three hours. Arrival time for critical faults is assumed to be one per 30 hours. Case study II is meant or represent densely deployed cells where SON functions

*Table 1. Model parameters for case studies*

| Parameter | Case Study: I | Case Study: II | Case Study: III |
|---|---|---|---|
| $\lambda_t$ hour$^{-1}$ | 1/8 | 1/3 | 1/8 |
| $\lambda_c$ hour$^{-1}$ | 1/80 | 1/30 | 1/80 |
| $\mu_{dc}$ hour$^{-1}$ | 1/6 | 1/6 | 1 |
| $\mu_c$ hour$^{-1}$ | 12 | 12 | 12 |
| Error | 0.00001 | 0.00001 | 0.00001 |

may need to be activated and deactivated more frequently, for example ultra-dense mmWave based deployment in 5G. In case study III, all parameters are assumed to be same as case study I except detection and compensation time that in case study I is assumed to be exponentially distributed with a mean value of 1 hour.

Transient Analysis of the three case studies is shown in Figure 3. For case study I, the probability of BS to be in optimal healthy state is around 95% after 1 hour and gradually decreases to around 60% after 24 hours period. There is a very low probability of the BS to be in outage state as critical failures rate is too small in our assumed model. For case study II, the probability of the network to be in sub-optimal state is 15% after 1 hour and it gradually increases to 62% after 24 hours since rate of arrival of trivial failures is high in case study II. In case study III, the probability of the BS to be in optimal state is around 88% after 24 hours. This indicates that decreased detection and compensation time has a profound effect on the network performance reliability. Therefore failures detection, diagnosis and compensation time should be as small as possible for achieving maximum performance. This calls for need for more agile self-healing methods in emerging cellular networks where increased complexity might cause higher fault arrival rate. The self-healing methods proposed in recent studies such as Zoha et al. (2015), Wang et al. (2014) and Lee (2013) are good candidates to overcome this problem. The occupancy time for the three case studies is show in in Figure 4. For case study I and III, the network remains in optimal state most of the time as compared to case study II in which sub-optimal time gradually increases with the passage of time. This is direct result of the

*Figure 3. Transient analysis of SON enabled BS for three case studies*

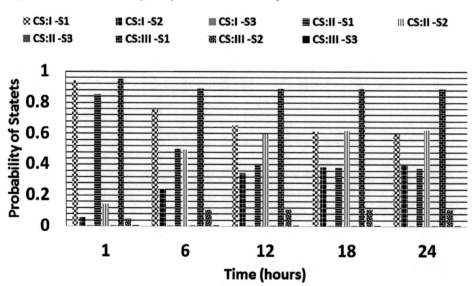

*Figure 4. Occupancy time of SON enabled BS for three case studies*

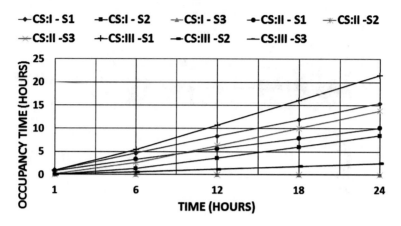

higher rate of arrival for trivial faults in case study II. The first passage times into state 2 and 3 is shown in Figure 5. The first passage time for the three case studies depends upon the mean arrival rate of trivial as well as critical failures so values of $\lambda_t$ and $\lambda_c$ both determine when a cell first experiences degradation and complete outage. As expected, the time to first experience sub-optimal performance is very small as compared to complete outage. First passage time is small in case study II as compared to the other two case studies due to higher rate of arrival of faults in case study II, compared to other two case studies. The limiting or steady state distribution is given in Figure 6. In the long run, a cell remains 58.3% and 88.9% of the time in optimal state for the case studies I and III respectively. However, for the case study II, it stays only 36.17% of the time in optimal state (63.77% in sub-optimal state) due to higher rate of trivial failures. The BS stays negligibly small amount of time in state 3 as critical failure rate is very small in our study.

## III. UTILITY OF THE DEVELOPED MODEL: FAULT PREDICTION FRAMEWORK (FPF)

Utilizing the analytical model developed in previous section, we propose Fault Predictive Framework (FPF) which predicts the occurrences of faults based upon the past database of failures (Figure 7). Historical data of past failures and misconfigurations of BS network parameters that occur routinely during operation of a cellular network can be utilized to estimate the $\lambda_t$, $\lambda_c$, $\mu_{dc}$ and $\mu_c$ parameters using standard machine learning tools. These estimated mean values then can be plugged into CTMC model and $Q$ and $R$ matrices can then be updated dynamically. The fitting of data to phase-

*Figure 5. First passage time of SON enabled BS for three Case Studies (CS)*

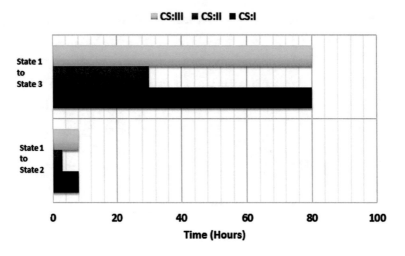

*Figure 6. Limiting (Steady State) distribution of SON enabled BS for three case studies*

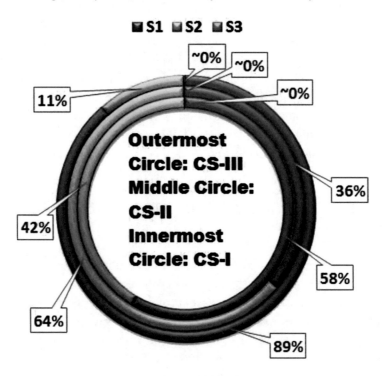

*Figure 7. Fault Prediction Framework (FPF)*

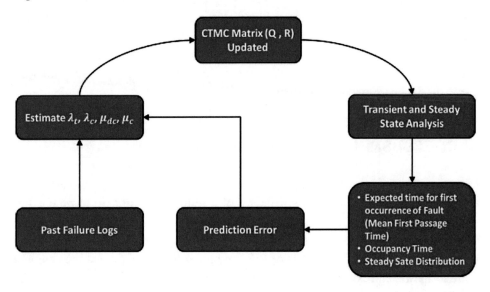

type distributions has been covered by various research studies, such as in Panchenko and Thümmler (2007). Based on updated $Q$ and $R$ matrices, Transient and Steady state analysis can then be run to compute new values for expected time for the first occurrence of fault, occupancy time and steady distribution. The difference between the predicted and actual values can be used to retrain the CTMC model parameters. In some of the cases, cell degradation is difficult to detect (Wang, Zhang & Zhang, 2013) e.g. in case of sleeping cells where no alarms are raised. In those cases, cell outage/degradation detection requires expensive site visits or drive testing that may take hours or days for the sub-optimal behavior to be detected. In majority of the cases, excessive customer complaints indicate the occurrence of faulty behavior of a cell. This results in significant reduction in quality of service and capacity. The probability of a cell to be sub-optimal at a given time period can be calculated by the proposed framework and can be exploited to minimize the degradation time. Once the predicted fault occurrence time is

near, prioritizing verification of each of the BS element or the configuration parameters can be initiated. Similarly, Occupancy time of the BS or Steady State distribution can be used as a KPI for cell performance. If the calculated values suggest that the time period which the cell will spend in sub-optimal or outage state is above some threshold value than that cell can be prioritized accordingly in the optimization process. The proposed framework can also aid in the diagnosis of the faults as this is one of the most difficult tasks faced by the BS subsystem engineers. If some record is maintained for the time interval of occurrence of fault and the corresponding root cause of that fault, as the expected suboptimal behavior or outage time approaches, the diagnosis should start right from the root cause already recorded in the table. This can result in significant reduction in diagnosis time and compensation time. The CTMC model and associated FPF framework presented in this paper thus can significantly improve reliability of network and provide enhanced user experience as expected from 5G.

## IV. CONCLUSION AND FUTURE WORK

In this paper, we presented a stochastic analytical model to analyze the effects of arrival of faults on the reliability behavior of a cellular network. Assuming exponential distributions for failures and recovery, a reliability model is developed using CTMC process. The proposed model, unlike previous studies on network reliability is not limited to structural aspects of BSs, and takes into account diverse potential fault scenarios and is capable to predict the expected time of first occurrence of the fault and long term reliability behavior of the BS. This model can adapt itself dynamically by learning from past database of network failures. Three different scenarios have been analyzed in terms of transient analysis, occupancy time, first passage time and the steady state distribution. As per the numerical results, mean arrival rate of trivial failures has profound effect on the reliability behavior of the cellular network. Another key finding is that, substantial gain in network reliability can be achieved by reducing BS's fault detection and recovery time, which strongly advocate the need for agile self-healing SON functions.

As for future work, the proposed model will be extended with non-exponential distribution for failures and recovery times. Moreover, methods will be developed to efficiently estimate CTMC model parameters by learning from the past failure logs collected from real network.

## ACKNOWLEGMENT

This work was made possible by NPRP grant No. 5-1047-2-437 from the Qatar National Research Fund (a member of The Qatar Foundation). The statements made herein are solely the responsibility of the authors. More information about this project can be found at www.qson.org.

# REFERENCES

Aliu, O. G., Imran, A., Imran, M. A., & Evans, B. (2013). A survey of self organisation in future cellular networks. *IEEE Communications Surveys and Tutorials, 15*(1), 336–361. doi:10.1109/SURV.2012.021312.00116

Dharmaraja, S., Jindal, V., & Varshney, U. (2008). Reliability and survivability analysis for UMTS networks: An analytical approach. *IEEE eTransactions on Network and Service Management, 5*(3), 132–142. doi:10.1109/TNSM.2009.031101

European Telecommunications Standards Institute. (2014). *Requirements for Further Advancements for Evolved Universal Terrestrial Radio Access*. Retrieved from http://www.etsi.org/deliver/etsi_tr/136900_136999/136913/10.00.00_60/tr_136913v100000p.pdf

Guida, M., Longo, M., & Postiglione, F. (2010, December). Performance evaluation of IMS-based core networks in presence of failures. In *Global Telecommunications Conference (GLOBECOM 2010)*, (pp. 1-5). IEEE. doi:10.1109/GLOCOM.2010.5683540

Hämäläinen, S., Sanneck, H., & Sartori, C. (2012). *LTE self-organising networks (SON): network management automation for operational efficiency*. John Wiley & Sons.

Imran, A., Zoha, A., & Abu-Dayya, A. (2014). Challenges in 5G: How to empower SON with big data for enabling 5G. *IEEE Network, 28*(6), 27–33. doi:10.1109/MNET.2014.6963801

Kulkarni, V. G. (2011). *Introduction to modelling and analysis of stochastic systems*. Springer.

Lateef, H. Y., Imran, A., & Abu-Dayya, A. (2013, September). A framework for classification of Self-Organising network conflicts and coordination algorithms. In *2013 IEEE 24th Annual International Symposium on Personal, Indoor, and Mobile Radio Communications (PIMRC)* (pp. 2898-2903). IEEE. doi:10.1109/PIMRC.2013.6666642

Lee, K., Lee, H., Jang, Y. U., & Cho, D. H. (2013). CoBRA: Cooperative beamforming-based resource allocation for self-healing in SON-based indoor mobile communication system. *IEEE Transactions on Wireless Communications, 12*(11), 5520–5528. doi:10.1109/TWC.2013.092013.121429

Osogami, T., & Harchol-Balter, M. (2006). Closed form solutions for mapping general distributions to quasi-minimal PH distributions. *Performance Evaluation, 63*(6), 524–552. doi:10.1016/j.peva.2005.06.002

Østerbø, O., & Grøndalen, O. (2012). Benefits of Self-Organizing Networks (SON) for mobile operators. *Journal of Computer Networks and Communications*.

Panchenko, A., & Thümmler, A. (2007). Efficient phase-type fitting with aggregated traffic traces. *Performance Evaluation*, *64*(7), 629–645. doi:10.1016/j.peva.2006.09.002

Pham, H. (2007). *System software reliability*. Springer Science & Business Media.

Reibman, A., & Trivedi, K. (1988). Numerical transient analysis of Markov models. *Computers & Operations Research*, *15*(1), 19–36. doi:10.1016/0305-0548(88)90026-3

Tipper, D., Charnsripinyo, C., Shin, H., & Dahlberg, T. (2002, January). Survivability analysis for mobile cellular networks.*Communication Networks and Distributed Systems Modeling and Simulation Conference*, 367-377.

Trivedi, K. S. (2008). *Probability & statistics with reliability, queuing and computer science applications*. John Wiley & Sons.

Wang, W., Liao, Q., & Zhang, Q. (2014). COD: A cooperative cell outage detection architecture for self-organizing femtocell networks. *IEEE Transactions on Wireless Communications*, *13*(11), 6007–6014. doi:10.1109/TWC.2014.2360865

Wang, W., Zhang, J., & Zhang, Q. (2013, April). Cooperative cell outage detection in self-organizing femtocell networks. In INFOCOM, 2013 Proceedings IEEE (pp. 782-790). IEEE. doi:10.1109/INFCOM.2013.6566865

Xie, L., Heegaard, P. E., & Jiang, Y. (2013, April). Network survivability under disaster propagation: Modeling and analysis. In 2013 IEEE Wireless Communications and Networking Conference (WCNC) (pp. 4730-4735). IEEE.

Zoha, A., Saeed, A., Imran, A., Imran, M. A., & Abu-Dayya, A. (2015, March). Data-driven analytics for automated cell outage detection in Self-Organizing Networks. In *Design of Reliable Communication Networks (DRCN), 2015 11th International Conference on the* (pp. 203-210). IEEE. doi:10.1109/DRCN.2015.7149014

# Chapter 10

# Spectral Efficiency Self-Optimization through Dynamic User Clustering and Beam Steering

**Md Salik Parwez**
*University of Oklahoma, USA*

**Ali Imran**
*University of Oklahoma, USA*

**Hasan Farooq**
*University of Oklahoma, USA*

**Hazem Refai**
*University of Oklahoma, USA*

## ABSTRACT

*This paper presents a novel scheme for spectral efficiency (SE) optimization through clustering of users. By clustering users with respect to their geographical concentration we propose a solution for dynamic steering of antenna beam, i.e., antenna azimuth and tilt optimization with respect to the most focal point in a cell that would maximize overall SE in the system. The proposed framework thus introduces the notion of elastic cells that can be potential component of 5G networks. The proposed scheme decomposes large-scale system-wide optimization problem into small-scale local sub-problems and thus provides a low complexity solution for dynamic system wide optimization. Every sub-problem involves clustering of users to determine focal point of the cell for given user distribution in time and space, and determining new values of azimuth and tilt that would optimize the overall system SE performance. To this end, we propose three user clustering algorithms to transform a given user distribution into the focal points that can be used in optimization; the first is based on received signal to interference ratio (SIR) at the user; the second is based on received signal level (RSL) at the user; the third and final one is based on relative*

DOI: 10.4018/978-1-5225-1750-4.ch010

*distances of users from the base stations. We also formulate and solve an optimization problem to determine optimal radii of clusters. The performances of proposed algorithms are evaluated through system level simulations. Performance comparison against benchmark where no elastic cell deployed, shows that a gain in spectral efficiency of up to 25% is possible depending upon user distribution in a cell.*

## I. INTRODUCTION

The tremendous increase in the number of mobile devices part of connectivity of anything to anything (also called Internet of Things (IoT)) and frequent emergence of diverse technologies are exerting extra pressure for dynamic data rate demand on wireless networks. Spectrum, which is regarded as one of the scarcest resources, must be efficiently utilized to meet those demands alongside the innovation and invention of new technologies and architectures. On one hand, there are a number of schemes being researched including, among others, Massive- Multiple Input Multiple Output (MIMO), Base Station (BS) densification, mmWave networks, and decoupled control and data plane architectures, that target the 5G and beyond networks to improve overall network efficiency. This paper, on the other hand, proposes to improve network spectral efficiency by optimizing the existing network parameters such as antenna azimuth and tilt angles, within the available resources. However, the diversity of users and their spatio-temporally varying requirements mandate the future networks to be not only heterogeneous and dense but also highly elastic. High network node density further increases the complexity to manage them. Hence, manual optimization of the network becomes highly challenging (Imran, Zoha & Abu-Dayya, 2014). Self-Organizing Networks (SON) has emerged as a technique to replace the manual handling by embedding intelligence and elasticity into the network (Aliu, Imran, Imran & Evans, 2013). SON enables the network to adapt to the changing environment by adjusting the network parameters autonomously. SON not only makes network highly efficient but also yields significant reduction in the network operational expenses (OPEX). In this article, we propose to optimize spectral efficiency (SE) by adaptively and simultaneously adjusting both antenna azimuth and tilt (i.e., in self-organizing manner) to steer the beam with respect to ever-changing user density and environment. To determine the highly dense regions of users within a cell, we propose and investigate three clustering algorithms, which when implemented, determine focal points in each cell. Once the clusters and their focal points are found, SE optimization algorithm is utilized to calculate new optimal azimuth and tilt values. Such online dynamic beam steering in real network could potentially be exploited using e.g., multi element antenna systems such as MIMO or massive MIMO which are being considered for emerging networks. The kind of beam steering proposed in this paper is much simpler and easier to implement as it does not require 1) tracking of individual users, 2) estimation of angle of arrival 3) estimation of channel. Instead the proposed solution requires simple antenna adjustments to change its azimuth and/or tilt by a few degrees as we will explain later.

The rest of the paper is organized as follows. Section II provides review of related work and outlines the novelty and contributions of this work. Section III presents system model while the self-optimization framework has been discussed in Section IV. Section V evaluates the performance using the numerical and simulation results. Section VI presents key conclusions.

## II. RELATED WORK

There are various techniques proposed in literature to enhance spectral efficiency by optimizing antenna parameters. In Seifi, Coldrey and Viberg (2012), user's average throughput has been maximized using BS-coordinated tilting. Authors in Muhammad, Abou-Jaoude, Hartmann and Mitschele-Thiel (2010), propose to adaptively adjust antenna tilt and pilot power to meet varying traffic load in the system. However, they take into consideration only the tilt and do not consider azimuth optimization. A similar solution is also proposed in Peyvandi, Imran, Imran and Tafazolli (2014) to study capacity and coverage optimization (CCO) use case of SON. However solution proposed in Peyvandi et al. (2014) optimizes the throughput of a single hotspot, it neither considers system-wide optimization nor does it address the dynamically changing user density throughout the cell. In Yilmaz, Hamalainen and Hamalainen (2009), Saur and Halbauer (2011), Halbauer and Saur (2012) and Seifi, Coldrey and Svensson (2012) switched beam tilting has been proposed, in which each BS utilizes one of the many pre-determined fixed tilts to maximize the users' throughput in certain region within a cell. In Coldrey and Svensson (2012), a framework has been proposed for dividing the cell into concentric region and applying switched beam tilting. However, solution in Coldrey and Svensson (2012) is studied in context of an isolated cell, and does not take into account interference from neighboring cells. In our recent work (Imran, Imran & Tafazolli, 2014), we proposed SON enabled system-wide SE optimization solutions for network with hotspots and relay stations. However, work in Imran et al. (2014) only considers tilt angle as the optimization parameter. Whereas in this article we propose self-optimization of both azimuth and tilt angle for changing user density. Second distinction of this work from Imran et al. (2014) is that, we present and compare three different user clustering algorithms to determine best representative point in a cell that can be used in the joint azimuth and tilt optimization processes. Furthermore, in this work we also present a method to determine optimal radius to cluster the users into groups. These contributions allow the optimization framework to be more user centric than that presented in Imran et al. (2014).

The significance of this work lies in the fact that joint optimization of both azimuth and tilt affectively paves the way for newly conceived cell-less deployment architecture—an architecture where cells won't have rigid foot prints. In such architecture active cells' shapes, sizes and numbers will vary with user distribution and demand (Imran et al., 2014). Such elastic cell-less architecture is one of the key features being envisioned for 5G. Proposed framework can be implemented in such elastic cell-less architecture by harnessing the beam steering capabilities of multi antenna element systems which are also a key component of 5G landscape.

The contribution of this paper is three fold. First, we propose a framework to optimize SE by adjusting antenna azimuth and tilt in self-organizing manner. Second, we propose and compare three algorithms to find focal points in each cell which can best represent given user distribution in the optimization process. This representation is manifested to reduce the computational complexity of the solution. In addition to that, we also formulate an optimization problem to determine the optimal radii within which the user density is the highest. These focal points and optimal radii provide us more accurate information about the location of the highest user density in the cell. We compare the results with the conventional fixed azimuth and tilt angle orientation. Results show that up to 25% gain in spectral efficiency can be achieved by using the proposed framework.

## III. SYSTEM MODEL

We consider a downlink transmission in multicellular network in which each BS has three sectors to begin with (which will be called hereafter 'cell'), each covering a span of 120 degree, as shown in Figure1. It is assumed that all user equipment (UEs) are outfitted with omnidirectional antenna with 0 dB gain. We use spectral efficiency (SE) in b/s/Hz as the optimization metric and we define it as the long term average bandwidth normalized throughput on a link given by $\log_2 (1 + SIR)$, where SIR stands for Signal to Interference Ratio.

Let $N$ denote the set of points corresponding to transmission antenna location of all sectors and $K$ denote the set of points representing the location of user in the system. The geometrical SIR perceived at a user location $k$ being served by $n^{th}$ sector can be given as:

$$\eta_k^n = \frac{P^n G_k^n \alpha \left(d_k^n\right)^{-\beta} \sigma_k^n}{\sum_{\forall m \in N \setminus n} \left(P^m G_k^m \alpha \left(d_k^m\right)^{-\beta} \sigma\right)} ; m, n \in N, k \in K \tag{1}$$

*Figure 1. Illustration of antenna tilt and azimuth angle*

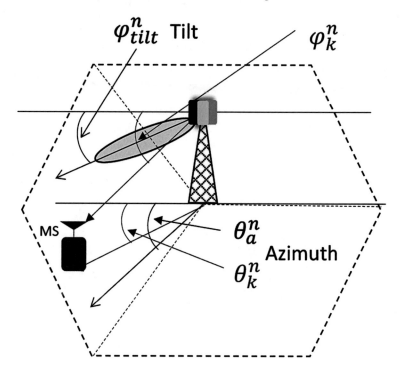

where $P^n$ indicates transmission power of the $n^{th}$ cell, is $d_k^n$ the distance between transmitting antenna location $n$ and UE location $k$. $\alpha$ and $\beta$ are pathloss model coefficient and exponents, respectively. $\sigma_k^n$ represents shadowing experienced by users at location $\sigma_k^n$ while receiving signal from $n^{th}$ transmitting antenna. $G_k^n$ represents antenna gain perceived at $k$ user location from $n^{th}$ antenna. As proposed by 3GPP (Viering, Dottling & Lobinger, 2009), the three dimensional antenna pattern can be given as

$$
G_k^n = 10^{0.1\left(\lambda_v\left(G_{\max}-\min\left(12\left(\frac{\varphi_k^n-\varphi_{tilt}^n}{B_v}\right)^2, A_{\max}\right)\right)+\lambda_h\left(G_{\max}-\min\left(12\left(\frac{\theta_k^n-\theta_a^n}{B_h}\right)^2, A_{\max}\right)\right)\right)}
\tag{2}
$$

and with simplification introduced in Viering et al. (2009), the above expression is reduced to

$$
G_k^n = 10^{-1.2\left(\lambda_v\left(\frac{\varphi_k^n-\varphi_{tilt}^n}{B_v}\right)^2+\lambda_h\left(\frac{\theta_k^n-\theta_a^n}{B_h}\right)^2\right)}
\tag{3}
$$

where $\varphi_k^n$ is the vertical angle at the $n^{th}$ BS in degrees from reference axis to the $k^{th}$ UE. $\varphi_{tilt}^n$ is the tilt angle of the $n^{th}$ cell as shown in Figure1. Also, $\theta_a^n$ represents azimuth angle orientation with respect to horizontal reference axis and $\theta_k^n$ is the angular distance of the $k^{th}$ user from horizontal reference axis. For simplicity, we use substitution in (1) as follows:

$$\delta_k^n = \sigma_k^n \alpha \left(d_k^n\right)^{-\beta}, \ \delta_k^m = \sigma_k^m \alpha \left(d_k^m\right)^{-\beta} \text{ and } \mu = \frac{-1.2\lambda_v}{B_v^{\ 2}}.$$

Using the above substitution and the gain from (3) into (1), we get the SIR at the UE represented as

$$\eta_k^n = \frac{\delta_k^n 10^{-1.2\left(\lambda_v\left(\frac{\varphi_k^n - \varphi_{tilt}^n}{B_v}\right)^2 + \lambda_h\left(\frac{\theta_k^n - \theta_a^n}{B_h}\right)^{22}\right)}}{\sum_{\forall m \in N \setminus n} \left(\delta_k^m 10^{-1.2\left(\lambda_v\left(\frac{\varphi_k^m - \varphi_{tilt}^m}{B_v}\right)^2 + \lambda_h\left(\frac{\theta_k^m - \theta_a^m}{B_h}\right)^2\right)}\right)} \tag{4}$$

where, $n, m \in N$; the indexes $n, m$ and $N$ represents the serving BS, interfering BS and the set of all the BSs respectively.

*Figure 2. Illustration of decomposition into triplets*

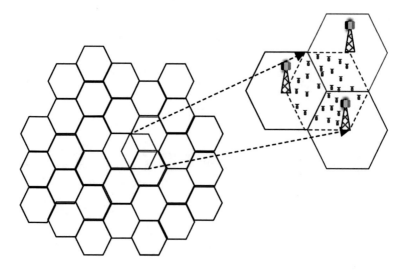

*Figure 3. Pseudo-code for SIR based clustering*

---

**Algorithm 1: SIR based Clustering**

---

Number of user in the cell: **N**, Radius of Cluster: **R**, User count: $C_i$
**for each cell do**
   calculate mean SIR of the cell: $SIR^{th}$
    **for each user do**
      calculate distance from other users $d_{ij}$ , $i \neq j$; $i, j = 1, \ldots N$
      **if** $d_{ij}$ <R *and* $SIR_i \geq SIR^{th}$ , **then**
       $C_i = C_i + 1$ ;
      **else**
       continue;
      **end**
    **end**
    select cluster with maximum user: **argmax($C_i, \ldots, C_N$)**
**end**

## IV. SELF ORGANIZING FRAMEWORK

In this section we detail the framework that decomposes the large scale problem into small scale problems to achieve distributed self-organizing solution. However, since the framework is built upon our recent work in Imran et al. (2014), we will describe it briefly. We then propose clustering algorithms to determine cell focal points. At the end of this section, we formulate problem to determine the optimal radius of clusters within the network cells.

## 1. Problem Formulation

As assumed, set $K$ represents the location of all the users in the system. So, the bandwidth normalized system throughput optimization can be expressed as (5) below.

$$\max_{\varphi_{tilt}^N, \theta_a^N} \hat{\xi} = \max_{\varphi_{tilt}^N, \theta_a^N} \sum_{\forall k \in K} \log_2 \left( 1 + \eta_k^n \left( \left( \varphi_{tilt}^N, \theta_a^N \right) \right) \right) \qquad (5)$$

It can be easily noticed that $\eta_k^n$ is a function of system wide azimuth and tilt angle. Since $K$ represents user and is usually a large number, which indicates that (5) is a large scale non-linear problem.

To overcome the difficulty of solving a large scale optimization problem, that will require real-time locations of all users in the system, we exploit the concept of determining a single focal point in each cell. The key attribute of such point is that it can affectively represent all the users in that cell during the optimization process. The validity of this approach was demonstrated in Lemma 1 in Imran et al. (2014) in context of tilt optimization. In this paper we extend that approach for joint azimuth and tilt approach and propose, in next section, three alternative algorithm to heuristically compute such single point for each cell.

Let $S$ denotes the set of all such focal points in all the cells. Then the objective function in (5) can be optimized only with respect to those focal points and thus can be approximated as

$$\max_{\varphi_{tilt}^{N},\theta_{a}^{N}} \sum_{\forall s \in S} log_2 \left( 1 + \eta_s^n \left( \left( \varphi_{tilt}^{N}, \theta_a^{N} \right) \right) \right) \tag{6}$$

Thus, compared to problem (5) which optimizes the antenna azimuth and tilt with respect to every user ' $K$ ' reduces to optimization problem (6) which optimizes with respect to only few focal points ' $S$ ' in the system, reducing the computational complexity.

## 2. Achieving a Distributed Solution

Equation (6), though has far less number of variables than (5), its solution still requires global coordination among all cells. A distributed solution can be enabled by exploiting the fact that in low power, small cell, high frequency band deployments being envisioned for 5G, interference will not propagate far beyond immediate neighbors. Thus the problem in (6) can be further approximated as

$$\max_{\varphi_{tilt}^{N},\theta_{a}^{N}} \hat{\xi} = \max_{\varphi_{tilt}^{N},\theta_{a}^{N}} \sum_{\forall s \in S} \log_2 \frac{1}{|N|} \left( 1 + \hat{\eta}_s^n \left( \left( \varphi_{tilt}^{N}, \theta_a^{N} \right) \right) \right) \tag{7}$$

where $\hat{\eta}_s^n$ represents the approximate SIR at the optimal point that considers observations from only its two immediate neighbors as shown in Figure 2. Lemma 1, corollary 3 in Imran et al. (2014), actually proves that for large $\beta$, $\hat{\xi}$ approaches the true value $\xi$. One particular scenario where beta is expected to be significantly large is mmWave based deployment (MacCartney, Zhang, Nie and Rappaport, 2013), which is being considered as an integral part of 5G landscape. Thus, according to the propositions 1 and 2 in Imran et al. (2014), (7) can be expressed as

$$\hat{\xi}_{N,\max} = \frac{1}{|N|} \sum_{\forall n \in N} \left\{ \max_{\varphi_{tilt}^{N},\theta_{a}^{N}} \frac{1}{|T_n|} \sum_{\forall s \in S_n} \log_2 \left( 1 + \hat{\eta}_s^n \left( \left( \varphi_{tilt}^{N}, \theta_a^{N} \right) \right) \right) \right\} \tag{8}$$

where, $S_n$ is the set of focal points in $n^{th}$ triplet and $|S_n| = |T_n| = T_n = 3, \forall n = N$.

Thus large scale centralized given in (6) is now reduced to small scale distributed optimization given in (8), for each triplet to be solved independently.

## 3. Throughput Optimization of Cluster in Each Cell

From the analysis presented above our problem in (5) has now been reduced to optimization of azimuth and tilt angle with respect to single focal points (to be determined) in individual triplets of cells. This optimization problem to be solved for every triplet is given as

$$\max \hat{\xi}\left(\varphi_{tilt}^1, \phi_a^1, \varphi_{tilt}^2, \theta_a^2, \varphi_{tilt}^3, \theta_a^3\right) \tag{9}$$

$$0 < \varphi_{tilt}^1, \varphi_{tilt}^2, \varphi_{tilt}^3 < 90 \tag{10}$$

$$0 < \theta_a^1 < 120 \tag{11}$$

$$121 < \theta_a^2 < 240 \tag{12}$$

$$241 < \theta_a^3 < 360 \tag{13}$$

Equation (14) presents the full form of objective function expressed in (9). Although the tilt angle varies from 0 to 90, in practice, the optimal value of tilt generally varies from 0 to 20 degree unless all users in the cell are concentrated at the base of base station. On the other hand in tri-sector system the optimal value of azimuth can be safely assumed to lie within ±15 degree of the nominal azimuth value unless user distribution is extremely skewed towards one edge of the sector, in which case it will be better to serve those users with that sector toward which user distribution is skewed. Our repeated computer simulations show that capping the range of azimuth adaptation is also necessary to limit inter-sector interference. These observations shortens the search space to

$$20 \times 20 \times 20 \times 15 \times 15 \times 15 = 27 \times 10^6.$$

This search space can be explored by any state of the art heuristic search algorithm that promises a global solution despite of non-convexity of the objective function. Noting that solution space, in this paper, is fairly small and well defined, we apply Simulated Annealing (SA) to explore the optimal azimuth and tilt and to make sure a global optimal within the curtailed search space is guaranteed.

## 4. Proposed Algorithms for Determining Cell Locus through Clustering

We investigate three different clustering algorithms to determine focal point (highly user dense region) in each cell which then can be used for azimuth and tilt optimization in real-time fashion. The key common idea behind the three clustering algorithms is that for every user in each cell, the number of users within a pre-specified radius will be determined. The users will be grouped into a cluster if they lie within that radius and if they meet the selected key performance indicators (KPI), including SIR, RSL and MS-BSs distance. The underlying KPI used to group users differentiates the three clustering algorithms. Users are grouped in clusters based on the selected KPI. The cluster which has the highest number of users is then chosen as the desired cluster. The mean (focal) point of the chosen cluster, for each cell in a triplet, is then determined. Those focal points are then used as the representative point with respect to which the optimization of antenna azimuth and tilt is performed in each triplet. In this study we assume cluster radius of 150 meter, but optimal radii of clusters will vary with cell radii, user distribution and propagation conditions.

*Figure 4. Pseudo-code for RSL based clustering*

**Algorithm 3: Distance Based Clustering**

```
Number of user in the cell: N, Radius of Cluster: R, User count: Ci
for each cell do
    for each user do
        calculate distance from other users dij , i≠j; i, j = 1,...N
        calculate distance from BSs: dserv, dint
        if dij <R and dserv < dint , then
            Ci= Ci +1 ;
        else
            continue;
        end
    end
    select cluster with maximum user: argmax(Ci,..., CN)
end
```

*Figure 5. Pseudo-code for distance based clustering*

---

**Algorithm 2: RSL Based Clustering**

---

Number of user in the cell: **N**;  Radius of Cluster: **R,** , User count: $C_i$

**for** each cell **do**

    **for** each user **do**

        calculate distance from other users $d_{ij}$ , $i \neq j$; $i, j = 1,...N$

        calculate RSL from BSs: $RSL_{serv}$, $RSL_{int}$

        **if** $d_{ij}$ <R *and* $RSL_{serv}$ < $RSL_{int}$ , **then**

            $C_i = C_i + 1$ ;

        **else**

           continue;

        **end**

    **end**

    select cluster with maximum user: **argmax($C_i$,..., $C_N$)**

**end**

---

The clustering algorithms based on different KPIs are described as follows:

1.  **Clustering Based on SIR:** Since users in each cell are served by their serving BS, other BSs in the triplet work as interferer. In this case, we first find the mean SIR of every cell, which acts as threshold SIR ($SIR_{th}$) for the users to be grouped into a cluster in addition to the condition that they should lie within the radius. In other words, to form the cluster around every user, the cluster users should be within the radius and their received SIR should be greater than or equal to $SIR_{th}$ of that cell under evaluation. Similar procedure is followed in others cells in the triplet, which will have their own $SIR_{th.}$ The pseudocode for the algorithm is given in Figure 3. Note that in emerging and futre cellular networks, with advent of location based services, accurate locations of indiviual users within a cell is known by the network. Given that location information is available at each base station, the SIR can be esitmated using (4). Alternatively, third generation partnership project (3GPP) Channel Quality inidicator (CQI) reported by users can also be exploited to estimate real time SIR. Thus, the clustering alrogirthms proposed in this paper are implementable in an online fashion with no additional signalling overhead.

$$\hat{\xi} = \log_2\left(1 + \frac{\delta_1^1 10^{-1.2\left(\lambda_v\left(\frac{\varphi_1^1-\varphi_{tilt}^1}{B_v}\right)^2+\lambda_h\left(\frac{\theta_1^1-\theta_a^1}{B_h}\right)^2\right)}}{\left(\delta_1^2 10^{-1.2\left(\lambda_v\left(\frac{\varphi_1^2-\varphi_{tilt}^2}{B_v}\right)^2+\lambda_h\left(\frac{\theta_1^2-\theta_a^2}{B_h}\right)^2\right)}\right) + \left(\delta_1^3 10^{-1.2\left(\lambda_v\left(\frac{\varphi_1^3-\varphi_{tilt}^3}{B_v}\right)^2+\lambda_h\left(\frac{\theta_1^3-\theta_a^3}{B_h}\right)^2\right)}\right)}\right)$$

$$+ \log_2\left(1 + \frac{\delta_2^2 10^{-1.2\left(\lambda_v\left(\frac{\varphi_2^2-\varphi_{tilt}^2}{B_v}\right)^2+\lambda_h\left(\frac{\theta_2^2-\theta_a^2}{B_h}\right)^2\right)}}{\left(\delta_2^1 10^{-1.2\left(\lambda_v\left(\frac{\varphi_2^1-\varphi_{tilt}^1}{B_v}\right)^2+\lambda_h\left(\frac{\theta_2^1-\theta_a^1}{B_h}\right)^2\right)}\right) + \left(\delta_2^3 10^{-1.2\left(\lambda_v\left(\frac{\varphi_2^3-\varphi_{tilt}^3}{B_v}\right)^2+\lambda_h\left(\frac{\theta_2^3-\theta_a^3}{B_h}\right)^2\right)}\right)}\right) \tag{14}$$

$$+ \log_2\left(1 + \frac{\delta_3^3 10^{-1.2\left(\lambda_v\left(\frac{\varphi_3^3-\varphi_{tilt}^3}{B_v}\right)^2+\lambda_h\left(\frac{\theta_3^3-\theta_a^3}{B_h}\right)^2\right)}}{\left(\delta_3^1 10^{-1.2\left(\lambda_v\left(\frac{\varphi_3^1-\varphi_{tilt}^1}{B_v}\right)^2+\lambda_h\left(\frac{\theta_3^1-\theta_a^1}{B_h}\right)^2\right)}\right) + \left(\delta_3^2 10^{-1.2\left(\lambda_v\left(\frac{\varphi_3^2-\varphi_{tilt}^2}{B_v}\right)^2+\lambda_h\left(\frac{\theta_3^2-\theta_a^2}{B_h}\right)^2\right)}\right)}\right)$$

2. **Clustering Based on RSL:** This method is similar to the one with SIR with the difference that SIR is replaced with RSL. The pseudocode for this algorithm is given in Figure 4 below. In emerging cellular neworks,, minmization of drive test (MDT), recently standardized by third 3GPP, contains RSL reports (Zoha, Saeed, Imran, Imran & Abu-Dayya, 2014). Thus standardization of MDT allows online implementaiton of the algorithm in Figure 4 without additional signalling overhead.

3. **Clustering Based on Distance from the BS:** This method also considers the highest number of user within the prespecified cluster radius with a center being the base station. However, the second criteria in this scheme is that the distance of the user from its serving BS is assumed to be smaller than the distance to the interfering BSs. The cluster formed using this method will ensure that it is nearer to the serving BS. This algorithm will be more useful to realize under

cell-less architecture where any user can connect to any BS depending upon signal availability. The pseudocode for this algorithm is given in Figure 5.

The choice of algorithm depends upon the motive behind clustering. For example, SIR based clustering will ensure that most of the users will be served with higher throughput most of the time. Similarly RSL based clustering can be chosen where throughput fairness is the major objective. While distance based clustering can be used for scenarios where there are no high rise buldings and towers that can work as interferers. Since their implementaiton, performance and complexity vary, there lies trade-off to be considered while selecting one scheme over the other. We also noticed that all three algorithms require the user density to be the highest but the location of the cluster will differ based on the selected KPI.

## 5. Problem Formulation for Optimal Radius

As discussed earlier the optimal threshold radii used in clustering algorithms will vary with a number of factors: cell size, user distribution, etc. Hence it is essential to determine the optimal radius. Furthermore, optimal radius calculation becomes more important while realizing cell-less architecture. In such cases, optimal radius would play key role in determining width of the beam directed toward each user.

While maximize spectral efficiency, we consider clustering based on SIR and the highest user density. Thus the spectral efficiency is a function of SIR at the user and total number of user (denoted by $N$) in the cell.

$$SE\left(\xi\right) = f\left(SIR\left(azimuth, tilt\right), N\right) \tag{15}$$

Although the radius may vary, we assume that optimal radius is employed and user distribution is assumed to have values of $50 < r < 150$.

If the radius of the hexagon is denoted by $R$, then the optimal radius can be determined by solving

$$\hat{r} = \max_{r}\left(\left(\pi r^2 \times \frac{2N}{3\sqrt{3R^2}}\right) \times SIR_n\right) \tag{16}$$

s.t.

$$50 < r < 150 \tag{17}$$

149

*Figure 6. CDF representation of SE achievable using CATO and SATO frameworks*

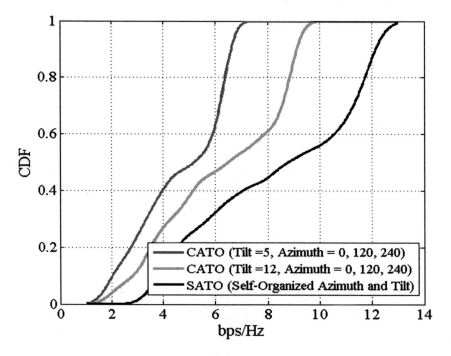

*Figure 7. Comparison of Bandwidth normalized throughput obtained using CATO and SATO*

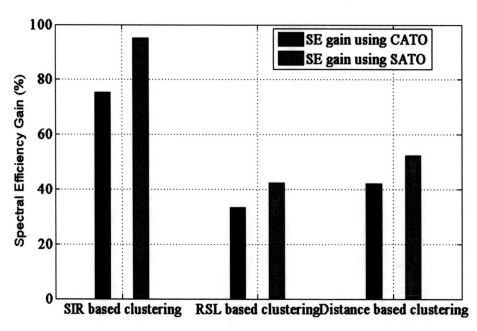

*Spectral Efficiency Self-Optimization*

*Table 1. Simulation parameters*

| Parameters | Values |
|---|---|
| System Topology | 19 BS with 3 sectors per BS |
| BS Transmission Power | 46 dBm |
| BS Inter site Distance | 500 meter |
| BS and UE height | 32 meter, 1.5 meter |
| UE Antenna Gain | 0 dB |
| Vertical Beamwidth | $70^0$ |
| Horizontal Beamwidth | $10^0$ |
| Vertical Gain Weight, $\lambda_v$ | 0.5 |
| Horizontal Gain Weight, $\lambda_h$ | 0.5 |
| Maximum Gain, $G_{max}$ | 18 dB |
| Maximum Attenuation, $A_{max}$ | 25 dB |
| Shadowing | 8 dB |
| Frequency | 2 GHz |
| Path Loss Model | 3GPP Urban Macro |

where $\hat{r}$ denotes the optimal radius of the cluster. It is important here to note that the cluster of range of radius in (17) is formed around every user and the radius satisfying (16) is chosen to be optimal. The $SIR_n$ indicates the SIR at each user $(n = 1,..,N)$. The above optimization problem is a differentiable and constrained problem; it can be easily solved through any nonlinear optimization algorithm.

# V. PERFORMANCE EVALUATION

In this section, we present numerical results using 3GPP recommended simulation parameters for LTE systems, given in Table 1.

To obtain the optimum azimuth and tilt, we applied heuristic search method 'Simulated Annealing' (Peyvandi, Imran, & Tafazolli). The detail of which is skipped because of limited space. We also implemented the optimization technique of sequential quadratic programming (SQP). Optimal azimuth and tilt were found using a brute force algorithm (SA) and the obtained results compared to those achieved using the SQP algorithm. The SQP results were confirmed and validated. We compared the bandwidth normalized throughput obtained by the proposed self-organized azimuth and tilt optimization framework, referred to as SATO herefoth, with those obtained using fixed azimuth and tilt which is referred to as centralized azimuth and tilt optimization (CATO) in Imran et al. (2014).

In Figure 6, we plot the SE commutative density function (CDF) achieved at the focal points using the proposed SATO framework. Then we compare the CDF against that obtained using CATO. We used fixed tilt of $5^o$ and $12^o$ and fixed typical azimuth of $0^o$, $120^o$, $240^o$ for cell 1, 2 and 3, respectively. We observe that at lower tilt of $5^o$, the performance is poor, because at smaller tilt, the beam is pointing towards the edge of the cell and hence exposed to higher interferences from other BSs. At the fixed tilt of $12^o$, and fixed regular azimuth, the performance improves as the interference from other BSs decreases. However, if the tilt goes on increasing the performance will degrade and limit the coverage at the cell edge. For uniform user distribution and for standard BS height and user end (UE) height, the fixed optimal tilt is centered around $12^o$ (Imran et al., 2014). We observe that using the

*Figure 8. SE gain using CATO and SATO for different focal points based on SIR clustering*

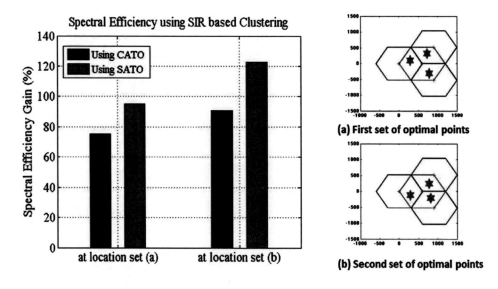

(a) First set of optimal points

(b) Second set of optimal points

SATO framework, SE gain of 1 to 3 bps/Hz is achievable over using the CATO framework. The SATO framework adjusts its azimuth and tilt automatically based on user density, thus optimizing the throughput at UE. It is clearly confirmed that the SATO technique outperforms the fixed scheme CATO.

Figure 7 presents the effect of different clustering techniques on SE performance, as the location of optimal focal points will vary for the investigated clustering schemes. It can be noticed that it is possible to achieve 10% to 25% gain in spectral efficiency, the highest being for the clustering based on SIR. This is because SIR based clustering considers not only the signal strength of the serving BS but also the interferences from other BSs. Other clustering algorithms although considers signal strength as well as the relative distance to the BS, they don't incorporate interference directly. Hence their performances are not as good as that of SIR based SATO. However, it's worth noting that the RSL and distance-based SATO are implementable with lower computational complexity thereby offering a trade-off between performance and complexity.

Moreover, for each algorithm the gain in SE varies depending upon the location of the focal points determined through clustering process. Figure 8 shows the variation in SE gain due to different focal point locations determined using SIR based clustering.

## VI. CONCLUSION

In this paper, we presented two important concepts implemented toward optimization of a system-wide SE. First, we discussed and analytically developed the framework to simultaneously configure two key antenna parameters (azimuth and tilt) in self-organizing manner. We decomposed the large-scale computationally taxing task into small-scale tasks by introducing the concept of triplet. Second, we determine a single focal point in each cell of a triplet that can be used to represent all the users in each cell. We evaluated three different clustering algorithms used to calculate the focal points that can best represent a given user distribution. These algorithms offer different levels of tradeoffs in implementation complexity and performance gain. The SIR based clustering algorithm offers the highest gain. We also proposed a method to determine optimal radius of a user cluster that further enables dynamic user centric optimization, which is essential to manage the continuously changing of user spatial distribution. We compared the SE gain against the conventional setting (fixed azimuth and tilt implementation). Obtained results confirm a 10% to 25% gain using the proposed scheme. The proposed framework is expandable towards cell-less deployment architecture in next generation network 5G, where cells are expected to adapt their sizes and shapes in user centric fashion by harnessing the flexibility inherent in multi element antenna systems. The proposed solution is also applicable to mmWave based systems as the interference in such system is mainly inter-sector interference and interference from neighboring cells can be neglected.

## ACKNOWLEDGMENT

This work was made possible by NPRP grant No. 5-1047-2-437 from the Qatar National Research Fund (a member of the Qatar Foundation). The statements made herein are solely the responsibility of authors.

## REFERENCES

Aliu, O. G., Imran, A., Imran, M. A., & Evans, B. (2013). A survey of self organisation in future cellular networks. *IEEE Communications Surveys and Tutorials*, *15*(1), 336–361. doi:10.1109/SURV.2012.021312.00116

Halbauer, H., Saur, S., Koppenborg, J., & Hoek, C. (2012, April). Interference avoidance with dynamic vertical beamsteering in real deployments. In Wireless Communications and Networking Conference Workshops (WCNCW), 2012 IEEE (pp. 294-299). IEEE. doi:10.1109/WCNCW.2012.6215509

Imran, A., Imran, M. A., Abu-Dayya, A., & Tafazolli, R. (2014). Self organization of tilts in relay enhanced networks: A distributed solution. *IEEE Transactions on Wireless Communications*, *13*(2), 764–779. doi:10.1109/TWC.2014.011614.130299

Imran, A., Imran, M. A., & Tafazolli, R. (2011, December). Distributed spectral efficiency optimization at hotspots through self organisation of BS tilts. In 2011 IEEE GLOBECOM Workshops (GC Wkshps) (pp. 570-574). IEEE. doi:10.1109/GLOCOMW.2011.6162515

Imran, A., Zoha, A., & Abu-Dayya, A. (2014). Challenges in 5G: How to empower SON with big data for enabling 5G. *IEEE Network*, *28*(6), 27–33. doi:10.1109/MNET.2014.6963801

Imran, M. A., Imran, A., & Tafazolli, R. (2011). Relay station access link spectral efficiency optimization through SO of macro BS tilts. *IEEE Communications Letters*, *15*(12), 1326–1328. doi:10.1109/LCOMM.2011.103111.1579

MacCartney, G. R., Zhang, J., Nie, S., & Rappaport, T. S. (2013, December). Path loss models for 5G millimeter wave propagation channels in urban microcells. In *2013 IEEE Global Communications Conference (GLOBECOM)*(pp. 3948-3953). IEEE. doi:10.1109/GLOCOM.2013.6831690

Muhammad, N. U. I., Abou-Jaoude, R., Hartmann, C., & Mitschele-Thiel, A. (2010, May). Self-Optimization of Antenna Tilt and Pilot Power for Dedicated Channels. WiOpt'10: Modeling and Optimization in Mobile, Ad Hoc, and Wireless Networks, 278-285.

Peyvandi, H., Imran, A., Imran, M. A., & Tafazolli, R. (2014, May). A target-following regime using Similarity Measure for Coverage and Capacity Optimization in Self-Organizing Cellular Networks with hot-spot. In *European Wireless 2014; 20th European Wireless Conference; Proceedings of* (pp. 1-6). VDE.

Peyvandi, H., Imran, M. A., & Tafazolli, R. (n.d.). *On Performance Optimization in Self-Organizing Networks using Enhanced Adaptive Simulated Annealing with Similarity Measure*. Academic Press.

Saur, S., & Halbauer, H. (2011, May). Exploring the vertical dimension of dynamic beam steering. In *Multi-Carrier Systems & Solutions (MC-SS), 2011 8th International Workshop on* (pp. 1-5). IEEE. doi:10.1109/MC-SS.2011.5910725

Seifi, N., Coldrey, M., & Svensson, T. (2012, December). Throughput optimization in MU-MIMO systems via exploiting BS antenna tilt. In 2012 IEEE Globecom Workshops (pp. 653-657). IEEE. doi:10.1109/GLOCOMW.2012.6477651

Seifi, N., Coldrey, M., & Viberg, M. (2012). Throughput optimization for MISO interference channels via coordinated user-specific tilting. *IEEE Communications Letters*, 16(8), 1248–1251. doi:10.1109/LCOMM.2012.060812.120756

Viering, I., Dottling, M., & Lobinger, A. (2009, June). A mathematical perspective of self-optimizing wireless networks. In *2009 IEEE International Conference on Communications* (pp. 1-6). IEEE. doi:10.1109/ICC.2009.5198628

Yilmaz, O. N., Hamalainen, S., & Hamalainen, J. (2009, September). System level analysis of vertical sectorization for 3GPP LTE. In *2009 6th International Symposium on Wireless Communication Systems* (pp. 453-457). IEEE.

Zoha, A., Saeed, A., Imran, A., Imran, M. A., & Abu-Dayya, A. (2014, September). A SON solution for sleeping cell detection using low-dimensional embedding of MDT measurements. In *2014 IEEE 25th Annual International Symposium on Personal, Indoor, and Mobile Radio Communication (PIMRC)* (pp. 1626-1630). IEEE. doi:10.1109/PIMRC.2014.7136428

# Related References

To continue our tradition of advancing information science and technology research, we have compiled a list of recommended IGI Global readings. These references will provide additional information and guidance to further enrich your knowledge and assist you with your own research and future publications.

Abramowicz, W., Stolarski, P., & Tomaszewski, T. (2013). Legal ontologies in ICT and law. In *Digital rights management: Concepts, methodologies, tools, and applications* (pp. 34–49). Hershey, PA: IGI Global. doi:10.4018/978-1-4666-2136-7.ch003

Adamich, T. (2012). Materials-to-standards alignment: How to "chunk" a whole cake and even use the "crumbs": State standards alignment models, learning objects, and formative assessment – methodologies and metadata for education. In L. Tomei (Ed.), *Advancing education with information communication technologies: Facilitating new trends* (pp. 165–178). Hershey, PA: IGI Global. doi:10.4018/978-1-61350-468-0.ch014

Adomi, E. E. (2011). Regulation of internet content. In E. Adomi (Ed.), *Frameworks for ICT policy: Government, social and legal issues* (pp. 233–246). Hershey, PA: IGI Global. doi:10.4018/978-1-61692-012-8.ch015

Aggestam, L. (2011). Guidelines for preparing organizations in developing countries for standards-based B2B. In *Global business: Concepts, methodologies, tools and applications* (pp. 206–228). Hershey, PA: IGI Global. doi:10.4018/978-1-60960-587-2.ch114

Akowuah, F., Yuan, X., Xu, J., & Wang, H. (2012). A survey of U.S. laws for health information security & privacy. *International Journal of Information Security and Privacy*, *6*(4), 40–54. doi:10.4018/jisp.2012100102

*Related References*

Akowuah, F., Yuan, X., Xu, J., & Wang, H. (2013). A survey of security standards applicable to health information systems. *International Journal of Information Security and Privacy*, *7*(4), 22–36. doi:10.4018/ijisp.2013100103

Al Hadid, I. (2012). Applying the certification's standards to the simulation study steps. In E. Abu-Taieh, A. El Sheikh, & M. Jafari (Eds.), *Technology engineering and management in aviation: Advancements and discoveries* (pp. 294–307). Hershey, PA: IGI Global. doi:10.4018/978-1-60960-887-3.ch017

Al Mohannadi, F., Arif, M., Aziz, Z., & Richardson, P. A. (2013). Adopting BIM standards for managing vision 2030 infrastructure development in Qatar. *International Journal of 3-D Information Modeling*, *2*(3), 64-73. doi:10.4018/ij3dim.2013070105

Al-Nuaimi, A. A. (2011). Using watermarking techniques to prove rightful ownership of web images. *International Journal of Information Technology and Web Engineering*, *6*(2), 29–39. doi:10.4018/jitwe.2011040103

Alejandre, G. M. (2013). IT security governance legal issues. In D. Mellado, L. Enrique Sánchez, E. Fernández-Medina, & M. Piattini (Eds.), *IT security governance innovations: Theory and research* (pp. 47–73). Hershey, PA: IGI Global. doi:10.4018/978-1-4666-2083-4.ch003

Alexandropoulou-Egyptiadou, E. (2013). The Hellenic framework for computer program copyright protection following the implementation of the relative european union directives. In *Digital rights management: Concepts, methodologies, tools, and applications* (pp. 738–745). Hershey, PA: IGI Global. doi:10.4018/978-1-4666-2136-7.ch033

Ali, S. (2012). Practical web application security audit following industry standards and compliance. In J. Zubairi & A. Mahboob (Eds.), *Cyber security standards, practices and industrial applications: Systems and methodologies* (pp. 259–279). Hershey, PA: IGI Global. doi:10.4018/978-1-60960-851-4.ch013

Alirezaee, M., & Afsharian, M. (2011). Measuring the effect of the rules and regulations on global malmquist index. *International Journal of Operations Research and Information Systems*, *2*(3), 64–78. doi:10.4018/joris.2011070105

Alirezaee, M., & Afsharian, M. (2013). Measuring the effect of the rules and regulations on global malmquist index. In J. Wang (Ed.), *Optimizing, innovating, and capitalizing on information systems for operations* (pp. 215–229). Hershey, PA: IGI Global. doi:10.4018/978-1-4666-2925-7.ch011

Alves de Lima, A., Carvalho dos Reis, P., Branco, J. C., Danieli, R., Osawa, C. C., Winter, E., & Santos, D. A. (2013). Scenario-patent protection compared to climate change: The case of green patents. *International Journal of Social Ecology and Sustainable Development*, 4(3), 61–70. doi:10.4018/jsesd.2013070105

Amirante, A., Castaldi, T., Miniero, L., & Romano, S. P. (2013). Protocol interactions among user agents, application servers, and media servers: Standardization efforts and open issues. In D. Kanellopoulos (Ed.), *Intelligent multimedia technologies for networking applications: Techniques and tools* (pp. 48–63). Hershey, PA: IGI Global. doi:10.4018/978-1-4666-2833-5.ch003

Anker, P. (2013). The impact of regulations on the business case for cognitive radio. In T. Lagkas, P. Sarigiannidis, M. Louta, & P. Chatzimisios (Eds.), *Evolution of cognitive networks and self-adaptive communication systems* (pp. 142–170). Hershey, PA: IGI Global. doi:10.4018/978-1-4666-4189-1.ch006

Antunes, A. M., Mendes, F. M., Schumacher, S. D., Quoniam, L., & Lima de Magalhães, J. (2014). The contribution of information science through intellectual property to innovation in the Brazilian health sector. In G. Jamil, A. Malheiro, & F. Ribeiro (Eds.), *Rethinking the conceptual base for new practical applications in information value and quality* (pp. 83–115). Hershey, PA: IGI Global. doi:10.4018/978-1-4666-4562-2.ch005

Atiskov, A. Y., Novikov, F. A., Fedorchenko, L. N., Vorobiev, V. I., & Moldovyan, N. A. (2013). Ontology-based analysis of cryptography standards and possibilities of their harmonization. In A. Elçi, J. Pieprzyk, A. Chefranov, M. Orgun, H. Wang, & R. Shankaran (Eds.), *Theory and practice of cryptography solutions for secure information systems* (pp. 1–33). Hershey, PA: IGI Global. doi:10.4018/978-1-4666-4030-6.ch001

Ayanso, A., & Herath, T. (2012). Law and technology at crossroads in cyberspace: Where do we go from here? In A. Dudley, J. Braman, & G. Vincenti (Eds.), *Investigating cyber law and cyber ethics: Issues, impacts and practices* (pp. 57–77). Hershey, PA: IGI Global. doi:10.4018/978-1-61350-132-0.ch004

Ayanso, A., & Herath, T. (2014). Law and technology at crossroads in cyberspace: Where do we go from here? In *Cyber behavior: Concepts, methodologies, tools, and applications* (pp. 1990–2010). Hershey, PA: IGI Global. doi:10.4018/978-1-4666-5942-1.ch105

### Related References

Aydogan-Duda, N. (2012). Branding innovation: The case study of Turkey. In N. Ekekwe & N. Islam (Eds.), *Disruptive technologies, innovation and global redesign: Emerging implications* (pp. 238–248). Hershey, PA: IGI Global. doi:10.4018/978-1-4666-0134-5.ch012

Bagby, J. W. (2011). Environmental standardization for sustainability. In Z. Luo (Ed.), *Green finance and sustainability: Environmentally-aware business models and technologies* (pp. 31–55). Hershey, PA: IGI Global. doi:10.4018/978-1-60960-531-5.ch002

Bagby, J. W. (2013). Insights from U.S. experience to guide international reliance on standardization: Achieving supply chain sustainability. *International Journal of Applied Logistics*, *4*(3), 25–46. doi:10.4018/jal.2013070103

Baggio, B., & Beldarrain, Y. (2011). Intellectual property in an age of open source and anonymity. In *Anonymity and learning in digitally mediated communications: Authenticity and trust in cyber education* (pp. 39–57). Hershey, PA: IGI Global. doi:10.4018/978-1-60960-543-8.ch003

Balzli, C. E., & Fragnière, E. (2012). How ERP systems are centralizing and standardizing the accounting function in public organizations for better and worse. In S. Chhabra & M. Kumar (Eds.), *Strategic enterprise resource planning models for e-government: Applications and methodologies* (pp. 55–72). Hershey, PA: IGI Global. doi:10.4018/978-1-60960-863-7.ch004

Banas, J. R. (2011). Standardized, flexible design of electronic learning environments to enhance learning efficiency and effectiveness. In A. Kitchenham (Ed.), *Models for interdisciplinary mobile learning: Delivering information to students* (pp. 66–86). Hershey, PA: IGI Global. doi:10.4018/978-1-60960-511-7.ch004

Bao, C., & Castresana, J. M. (2011). Interoperability approach in e-learning standardization processes. In F. Lazarinis, S. Green, & E. Pearson (Eds.), *Handbook of research on e-learning standards and interoperability: Frameworks and issues* (pp. 399–418). Hershey, PA: IGI Global. doi:10.4018/978-1-61692-789-9.ch020

Bao, C., & Castresana, J. M. (2012). Interoperability approach in e-learning standardization processes. In *Virtual learning environments: Concepts, methodologies, tools and applications* (pp. 542–560). Hershey, PA: IGI Global. doi:10.4018/978-1-4666-0011-9.ch307

Barrett, B. (2011). Evaluating and implementing teaching standards: Providing quality online teaching strategies and techniques standards. In F. Lazarinis, S. Green, & E. Pearson (Eds.), *Developing and utilizing e-learning applications* (pp. 66–83). Hershey, PA: IGI Global. doi:10.4018/978-1-61692-791-2.ch004

Berleur, J. (2011). Ethical and social issues of the internet governance regulations. In D. Haftor & A. Mirijamdotter (Eds.), *Information and communication technologies, society and human beings: Theory and framework (festschrift in honor of Gunilla Bradley)* (pp. 466–476). Hershey, PA: IGI Global. doi:10.4018/978-1-60960-057-0.ch038

Bhattathiripad, V. P. (2014). Software copyright infringement and litigation. In *Judiciary-friendly forensics of software copyright infringement* (pp. 35–55). Hershey, PA: IGI Global. doi:10.4018/978-1-4666-5804-2.ch002

Bin, X., & Chuan, T. K. (2011). The effect of business characteristics on the methods of knowledge protections. *International Journal of Social Ecology and Sustainable Development*, 2(3), 34–60. doi:10.4018/jsesd.2011070103

Bin, X., & Chuan, T. K. (2013). The effect of business characteristics on the methods of knowledge protections. In E. Carayannis (Ed.), *Creating a sustainable ecology using technology-driven solutions* (pp. 172–200). Hershey, PA: IGI Global. doi:10.4018/978-1-4666-3613-2.ch013

Bin, X., & Chuan, T. K. (2013). The effect of business characteristics on the methods of knowledge protections. In *Digital rights management: Concepts, methodologies, tools, and applications* (pp. 1283–1311). Hershey, PA: IGI Global. doi:10.4018/978-1-4666-2136-7.ch063

Bogers, M., Bekkers, R., & Granstrand, O. (2012). Intellectual property and licensing strategies in open collaborative innovation. In C. de Pablos Heredero & D. López (Eds.), *Open innovation in firms and public administrations: Technologies for value creation* (pp. 37–58). Hershey, PA: IGI Global. doi:10.4018/978-1-61350-341-6.ch003

Bogers, M., Bekkers, R., & Granstrand, O. (2013). Intellectual property and licensing strategies in open collaborative innovation. In *Digital rights management: Concepts, methodologies, tools, and applications* (pp. 1204–1224). Hershey, PA: IGI Global. doi:10.4018/978-1-4666-2136-7.ch059

*Related References*

Bourcier, D. (2013). Law and governance: The genesis of the commons. In F. Doridot, P. Duquenoy, P. Goujon, A. Kurt, S. Lavelle, N. Patrignani, & A. Santuccio et al. (Eds.), *Ethical governance of emerging technologies development* (pp. 166–183). Hershey, PA: IGI Global. doi:10.4018/978-1-4666-3670-5.ch011

Bousquet, F., Fomin, V. V., & Drillon, D. (2011). Anticipatory standards development and competitive intelligence. *International Journal of Business Intelligence Research*, 2(1), 16–30. doi:10.4018/jbir.2011010102

Bousquet, F., Fomin, V. V., & Drillon, D. (2013). Anticipatory standards development and competitive intelligence. In R. Herschel (Ed.), *Principles and applications of business intelligence research* (pp. 17–30). Hershey, PA: IGI Global. doi:10.4018/978-1-4666-2650-8.ch002

Brabazon, A. (2013). Optimal patent design: An agent-based modeling approach. In B. Alexandrova-Kabadjova, S. Martinez-Jaramillo, A. Garcia-Almanza, & E. Tsang (Eds.), *Simulation in computational finance and economics: Tools and emerging applications* (pp. 280–302). Hershey, PA: IGI Global. doi:10.4018/978-1-4666-2011-7.ch014

Bracci, F., Corradi, A., & Foschini, L. (2014). Cloud standards: Security and interoperability issues. In H. Mouftah & B. Kantarci (Eds.), *Communication infrastructures for cloud computing* (pp. 465–495). Hershey, PA: IGI Global. doi:10.4018/978-1-4666-4522-6.ch020

Briscoe, D. R. (2012). Globalization and international labor standards, codes of conduct, and ethics: An International HRM perspective. In C. Wankel & S. Malleck (Eds.), *Ethical models and applications of globalization: Cultural, socio-political and economic perspectives* (pp. 1–22). Hershey, PA: IGI Global. doi:10.4018/978-1-61350-332-4.ch001

Briscoe, D. R. (2014). Globalization and international labor standards, codes of conduct, and ethics: An International HRM perspective. In *Cross-cultural interaction: Concepts, methodologies, tools and applications* (pp. 40–62). Hershey, PA: IGI Global. doi:10.4018/978-1-4666-4979-8.ch004

Brooks, R. G., & Geradin, D. (2011). Interpreting and enforcing the voluntary FRAND commitment. *International Journal of IT Standards and Standardization Research*, 9(1), 1–23. doi:10.4018/jitsr.2011010101

Brown, C. A. (2013). Common core state standards: The promise for college and career ready students in the U.S. In V. Wang (Ed.), *Handbook of research on teaching and learning in K-20 education* (pp. 50–82). Hershey, PA: IGI Global. doi:10.4018/978-1-4666-4249-2.ch004

Buyurgan, N., Rardin, R. L., Jayaraman, R., Varghese, V. M., & Burbano, A. (2011). A novel GS1 data standard adoption roadmap for healthcare providers. *International Journal of Healthcare Information Systems and Informatics*, *6*(4), 42–59. doi:10.4018/jhisi.2011100103

Buyurgan, N., Rardin, R. L., Jayaraman, R., Varghese, V. M., & Burbano, A. (2013). A novel GS1 data standard adoption roadmap for healthcare providers. In J. Tan (Ed.), *Healthcare information technology innovation and sustainability: Frontiers and adoption* (pp. 41–57). Hershey, PA: IGI Global. doi:10.4018/978-1-4666-2797-0.ch003

Campolo, C., Cozzetti, H. A., Molinaro, A., & Scopigno, R. M. (2012). PHY/MAC layer design in vehicular ad hoc networks: Challenges, standard approaches, and alternative solutions. In R. Aquino-Santos, A. Edwards, & V. Rangel-Licea (Eds.), *Wireless technologies in vehicular ad hoc networks: Present and future challenges* (pp. 70–100). Hershey, PA: IGI Global. doi:10.4018/978-1-4666-0209-0.ch004

Cantatore, F. (2014). Copyright support structures. In *Authors, copyright, and publishing in the digital era* (pp. 81–93). Hershey, PA: IGI Global. doi:10.4018/978-1-4666-5214-9.ch005

Cantatore, F. (2014). History and development of copyright. In *Authors, copyright, and publishing in the digital era* (pp. 10–32). Hershey, PA: IGI Global. doi:10.4018/978-1-4666-5214-9.ch002

Cantatore, F. (2014). Research findings: Authors' perceptions and the copyright framework. In Authors, copyright, and publishing in the digital era (pp. 147-189). Hershey, PA: IGI Global. doi:10.4018/978-1-4666-5214-9.ch008

Cassini, J., Medlin, B. D., & Romaniello, A. (2011). Forty years of federal legislation in the area of data protection and information security. In H. Nemati (Ed.), *Pervasive information security and privacy developments: Trends and advancements* (pp. 14–23). Hershey, PA: IGI Global. doi:10.4018/978-1-61692-000-5.ch002

*Related References*

Charlesworth, A. (2012). Addressing legal issues in online research, publication and archiving: A UK perspective. In C. Silva (Ed.), *Online research methods in urban and planning studies: Design and outcomes* (pp. 368–393). Hershey, PA: IGI Global. doi:10.4018/978-1-4666-0074-4.ch022

Chaudhary, C., & Kang, I. S. (2011). Pirates of the copyright and cyberspace: Issues involved. In R. Santanam, M. Sethumadhavan, & M. Virendra (Eds.), *Cyber security, cyber crime and cyber forensics: Applications and perspectives* (pp. 59–68). Hershey, PA: IGI Global. doi:10.4018/978-1-60960-123-2.ch005

Chen, L., Hu, W., Yang, M., & Zhang, L. (2011). Security and privacy issues in secure e-mail standards and services. In H. Nemati (Ed.), *Security and privacy assurance in advancing technologies: New developments* (pp. 174–185). Hershey, PA: IGI Global. doi:10.4018/978-1-60960-200-0.ch013

Ciaghi, A., & Villafiorita, A. (2012). Law modeling and BPR for public administration improvement. In K. Bwalya & S. Zulu (Eds.), *Handbook of research on e-government in emerging economies: Adoption, E-participation, and legal frameworks* (pp. 391–410). Hershey, PA: IGI Global. doi:10.4018/978-1-4666-0324-0.ch019

Ciptasari, R. W., & Sakurai, K. (2013). Multimedia copyright protection scheme based on the direct feature-based method. In K. Kondo (Ed.), *Multimedia information hiding technologies and methodologies for controlling data* (pp. 412–439). Hershey, PA: IGI Global. doi:10.4018/978-1-4666-2217-3.ch019

Clark, L. A., Jones, D. L., & Clark, W. J. (2012). Technology innovation and the policy vacuum: A call for ethics, norms, and laws to fill the void. *International Journal of Technoethics*, *3*(1), 1–13. doi:10.4018/jte.2012010101

Cooklev, T. (2013). The role of standards in engineering education. In K. Jakobs (Ed.), *Innovations in organizational IT specification and standards development* (pp. 129–137). Hershey, PA: IGI Global. doi:10.4018/978-1-4666-2160-2.ch007

Cooper, A. R. (2013). Key challenges in the design of learning technology standards: Observations and proposals. In K. Jakobs (Ed.), *Innovations in organizational IT specification and standards development* (pp. 241–249). Hershey, PA: IGI Global. doi:10.4018/978-1-4666-2160-2.ch014

Cordella, A. (2011). Emerging standardization. *International Journal of Actor-Network Theory and Technological Innovation*, *3*(3), 49–64. doi:10.4018/jantti.2011070104

Cordella, A. (2013). Emerging standardization. In A. Tatnall (Ed.), *Social and professional applications of actor-network theory for technology development* (pp. 221–237). Hershey, PA: IGI Global. doi:10.4018/978-1-4666-2166-4.ch017

Curran, K., & Lautman, R. (2011). The problems of jurisdiction on the internet. *International Journal of Ambient Computing and Intelligence*, *3*(3), 36–42. doi:10.4018/jaci.2011070105

Dani, D. E., Salloum, S., Khishfe, R., & BouJaoude, S. (2013). A tool for analyzing science standards and curricula for 21st century science education. In M. Khine, & I. Saleh (Eds.), *Approaches and strategies in next generation science learning* (pp. 265-289). Hershey, PA: IGI Global. doi:10.4018/978-1-4666-2809-0.ch014

De Silva, S. (2012). Legal issues with FOS-ERP: A UK law perspective. In R. Atem de Carvalho & B. Johansson (Eds.), *Free and open source enterprise resource planning: Systems and strategies* (pp. 102–115). Hershey, PA: IGI Global. doi:10.4018/978-1-61350-486-4.ch007

de Vries, H. J. (2011). Implementing standardization education at the national level. *International Journal of IT Standards and Standardization Research*, *9*(2), 72–83. doi:10.4018/jitsr.2011070104

de Vries, H. J. (2013). Implementing standardization education at the national level. In K. Jakobs (Ed.), *Innovations in organizational IT specification and standards development* (pp. 116–128). Hershey, PA: IGI Global. doi:10.4018/978-1-4666-2160-2.ch006

de Vuyst, B., & Fairchild, A. (2012). Legal and economic justification for software protection. *International Journal of Open Source Software and Processes*, *4*(3), 1–12. doi:10.4018/ijossp.2012070101

Dedeke, A. (2012). Politics hinders open standards in the public sector: The Massachusetts open document format decision. In C. Reddick (Ed.), *Cases on public information management and e-government adoption* (pp. 1–23). Hershey, PA: IGI Global. doi:10.4018/978-1-4666-0981-5.ch001

Delfmann, P., Herwig, S., Lis, L., & Becker, J. (2012). Supporting conceptual model analysis using semantic standardization and structural pattern matching. In S. Smolnik, F. Teuteberg, & O. Thomas (Eds.), *Semantic technologies for business and information systems engineering: Concepts and applications* (pp. 125–149). Hershey, PA: IGI Global. doi:10.4018/978-1-60960-126-3.ch007

*Related References*

den Uijl, S., de Vries, H. J., & Bayramoglu, D. (2013). The rise of MP3 as the market standard: How compressed audio files became the dominant music format. *International Journal of IT Standards and Standardization Research, 11*(1), 1–26. doi:10.4018/jitsr.2013010101

Dickerson, J., & Coleman, H. V. (2012). Technology, e-leadership and educational administration in schools: Integrating standards with context and guiding questions. In V. Wang (Ed.), *Encyclopedia of e-leadership, counseling and training* (pp. 408–422). Hershey, PA: IGI Global. doi:10.4018/978-1-61350-068-2.ch030

Dindaroglu, B. (2013). R&D productivity and firm size in semiconductors and pharmaceuticals: Evidence from citation yields. In I. Yetkiner, M. Pamukcu, & E. Erdil (Eds.), *Industrial dynamics, innovation policy, and economic growth through technological advancements* (pp. 92–113). Hershey, PA: IGI Global. doi:10.4018/978-1-4666-1978-4.ch006

Ding, W. (2011). Development of intellectual property of communications enterprise and analysis of current situation of patents in emerging technology field. *International Journal of Advanced Pervasive and Ubiquitous Computing, 3*(2), 21–28. doi:10.4018/japuc.2011040103

Ding, W. (2013). Development of intellectual property of communications enterprise and analysis of current situation of patents in emerging technology field. In T. Gao (Ed.), *Global applications of pervasive and ubiquitous computing* (pp. 89–96). Hershey, PA: IGI Global. doi:10.4018/978-1-4666-2645-4.ch010

Dorloff, F., & Kajan, E. (2012). Balancing of heterogeneity and interoperability in e-business networks: The role of standards and protocols. *International Journal of E-Business Research, 8*(4), 15–33. doi:10.4018/jebr.2012100102

Dorloff, F., & Kajan, E. (2012). Efficient and interoperable e-business –Based on frameworks, standards and protocols: An introduction. In E. Kajan, F. Dorloff, & I. Bedini (Eds.), *Handbook of research on e-business standards and protocols: Documents, data and advanced web technologies* (pp. 1–20). Hershey, PA: IGI Global. doi:10.4018/978-1-4666-0146-8.ch001

Driouchi, A., & Kadiri, M. (2013). Challenges to intellectual property rights from information and communication technologies, nanotechnologies and microelectronics. In *Digital rights management: Concepts, methodologies, tools, and applications* (pp. 1474–1492). Hershey, PA: IGI Global. doi:10.4018/978-1-4666-2136-7.ch075

Dubey, M., & Hirwade, M. (2013). Copyright relevancy at stake in libraries of the digital era. In T. Ashraf & P. Gulati (Eds.), *Design, development, and management of resources for digital library services* (pp. 379–384). Hershey, PA: IGI Global. doi:10.4018/978-1-4666-2500-6.ch030

Egyedi, T. M. (2011). Between supply and demand: Coping with the impact of standards change. In *Global business: Concepts, methodologies, tools and applications* (pp. 105–120). Hershey, PA: IGI Global. doi:10.4018/978-1-60960-587-2.ch108

Egyedi, T. M., & Koppenhol, A. (2013). The standards war between ODF and OOXML: Does competition between overlapping ISO standards lead to innovation? In K. Jakobs (Ed.), *Innovations in organizational IT specification and standards development* (pp. 79–90). Hershey, PA: IGI Global. doi:10.4018/978-1-4666-2160-2.ch004

Egyedi, T. M., & Muto, S. (2012). Standards for ICT: A green strategy in a grey sector. *International Journal of IT Standards and Standardization Research*, *10*(1), 34–47. doi:10.4018/jitsr.2012010103

El Kharbili, M., & Pulvermueller, E. (2012). Semantic policies for modeling regulatory process compliance. In S. Smolnik, F. Teuteberg, & O. Thomas (Eds.), *Semantic technologies for business and information systems engineering: Concepts and applications* (pp. 311–336). Hershey, PA: IGI Global. doi:10.4018/978-1-60960-126-3.ch016

El Kharbili, M., & Pulvermueller, E. (2013). Semantic policies for modeling regulatory process compliance. In *IT policy and ethics: Concepts, methodologies, tools, and applications* (pp. 218–243). Hershey, PA: IGI Global. doi:10.4018/978-1-4666-2919-6.ch011

Ervin, K. (2014). Legal and ethical considerations in the implementation of electronic health records. In J. Krueger (Ed.), *Cases on electronic records and resource management implementation in diverse environments* (pp. 193–210). Hershey, PA: IGI Global. doi:10.4018/978-1-4666-4466-3.ch012

Escayola, J., Trigo, J., Martínez, I., Martínez-Espronceda, M., Aragüés, A., Sancho, D., & García, J. et al. (2012). Overview of the ISO/ieee11073 family of standards and their applications to health monitoring. In W. Chen, S. Oetomo, & L. Feijs (Eds.), *Neonatal monitoring technologies: Design for integrated solutions* (pp. 148–173). Hershey, PA: IGI Global. doi:10.4018/978-1-4666-0975-4.ch007

### Related References

Escayola, J., Trigo, J., Martínez, I., Martínez-Espronceda, M., Aragüés, A., Sancho, D., . . . García, J. (2013). Overview of the ISO/IEEE11073 family of standards and their applications to health monitoring. In User-driven healthcare: Concepts, methodologies, tools, and applications (pp. 357-381). Hershey, PA: IGI Global. doi:10.4018/978-1-4666-2770-3.ch018

Espada, J. P., Martínez, O. S., García-Bustelo, B. C., Lovelle, J. M., & Ordóñez de Pablos, P. (2011). Standardization of virtual objects. In M. Lytras, P. Ordóñez de Pablos, & E. Damiani (Eds.), *Semantic web personalization and context awareness: Management of personal identities and social networking* (pp. 7–21). Hershey, PA: IGI Global. doi:10.4018/978-1-61520-921-7.ch002

Falkner, N. J. (2011). Security technologies and policies in organisations. In M. Quigley (Ed.), *ICT ethics and security in the 21st century: New developments and applications* (pp. 196–213). Hershey, PA: IGI Global. doi:10.4018/978-1-60960-573-5.ch010

Ferrer-Roca, O. (2011). Standards in telemedicine. In A. Moumtzoglou & A. Kastania (Eds.), *E-health systems quality and reliability: Models and standards* (pp. 220–243). Hershey, PA: IGI Global. doi:10.4018/978-1-61692-843-8.ch017

Ferullo, D. L., & Soules, A. (2012). Managing copyright in a digital world. *International Journal of Digital Library Systems, 3*(4), 1–25. doi:10.4018/ijdls.2012100101

Fichtner, J. R., & Simpson, L. A. (2011). Legal issues facing companies with products in a digital format. In T. Strader (Ed.), *Digital product management, technology and practice: Interdisciplinary perspectives* (pp. 32–52). Hershey, PA: IGI Global. doi:10.4018/978-1-61692-877-3.ch003

Fichtner, J. R., & Simpson, L. A. (2013). Legal issues facing companies with products in a digital format. In *Digital rights management: Concepts, methodologies, tools, and applications* (pp. 1334–1354). Hershey, PA: IGI Global. doi:10.4018/978-1-4666-2136-7.ch066

Folmer, E. (2012). BOMOS: Management and development model for open standards. In E. Kajan, F. Dorloff, & I. Bedini (Eds.), *Handbook of research on e-business standards and protocols: Documents, data and advanced web technologies* (pp. 102–128). Hershey, PA: IGI Global. doi:10.4018/978-1-4666-0146-8.ch006

Fomin, V. V. (2012). Standards as hybrids: An essay on tensions and juxtapositions in contemporary standardization. *International Journal of IT Standards and Standardization Research, 10*(2), 59–68. doi:10.4018/jitsr.2012070105

Fomin, V. V., & Matinmikko, M. (2014). The role of standards in the development of new informational infrastructure. In M. Khosrow-Pour (Ed.), *Systems and software development, modeling, and analysis: New perspectives and methodologies* (pp. 149–160). Hershey, PA: IGI Global. doi:10.4018/978-1-4666-6098-4.ch006

Fomin, V. V., Medeisis, A., & Vitkute-Adžgauskiene, D. (2012). Pre-standardization of cognitive radio systems. *International Journal of IT Standards and Standardization Research, 10*(1), 1–16. doi:10.4018/jitsr.2012010101

Francia, G., & Hutchinson, F. S. (2012). Regulatory and policy compliance with regard to identity theft prevention, detection, and response. In T. Chou (Ed.), *Information assurance and security technologies for risk assessment and threat management: Advances* (pp. 292–322). Hershey, PA: IGI Global. doi:10.4018/978-1-61350-507-6.ch012

Francia, G. A., & Hutchinson, F. S. (2014). Regulatory and policy compliance with regard to identity theft prevention, detection, and response. In *Crisis management: Concepts, methodologies, tools and applications* (pp. 280–310). Hershey, PA: IGI Global. doi:10.4018/978-1-4666-4707-7.ch012

Fulkerson, D. M. (2012). Copyright. In D. Fulkerson (Ed.), *Remote access technologies for library collections: Tools for library users and managers* (pp. 33–48). Hershey, PA: IGI Global. doi:10.4018/978-1-4666-0234-2.ch003

Galinski, C., & Beckmann, H. (2014). Concepts for enhancing content quality and eaccessibility: In general and in the field of eprocurement. In *Assistive technologies: Concepts, methodologies, tools, and applications* (pp. 180–197). Hershey, PA: IGI Global. doi:10.4018/978-1-4666-4422-9.ch010

Gaur, R. (2013). Facilitating access to Indian cultural heritage: Copyright, permission rights and ownership issues vis-à-vis IGNCA collections. In *Digital rights management: Concepts, methodologies, tools, and applications* (pp. 817–833). Hershey, PA: IGI Global. doi:10.4018/978-1-4666-2136-7.ch038

Geiger, C. (2011). Copyright and digital libraries: Securing access to information in the digital age. In I. Iglezakis, T. Synodinou, & S. Kapidakis (Eds.), *E-publishing and digital libraries: Legal and organizational issues* (pp. 257–272). Hershey, PA: IGI Global. doi:10.4018/978-1-60960-031-0.ch013

Geiger, C. (2013). Copyright and digital libraries: Securing access to information in the digital age. In *Digital rights management: Concepts, methodologies, tools, and applications* (pp. 99–114). Hershey, PA: IGI Global. doi:10.4018/978-1-4666-2136-7.ch007

**Related References**

Gencer, M. (2012). The evolution of IETF standards and their production. *International Journal of IT Standards and Standardization Research*, *10*(1), 17–33. doi:10.4018/jitsr.2012010102

Gillam, L., & Vartapetiance, A. (2014). Gambling with laws and ethics in cyberspace. In R. Luppicini (Ed.), *Evolving issues surrounding technoethics and society in the digital age* (pp. 149–170). Hershey, PA: IGI Global. doi:10.4018/978-1-4666-6122-6.ch010

Grandinetti, L., Pisacane, O., & Sheikhalishahi, M. (2014). Standardization. In *Pervasive cloud computing technologies: Future outlooks and interdisciplinary perspectives* (pp. 75–96). Hershey, PA: IGI Global. doi:10.4018/978-1-4666-4683-4.ch004

Grant, S., & Young, R. (2013). Concepts and standardization in areas relating to competence. In K. Jakobs (Ed.), *Innovations in organizational IT specification and standards development* (pp. 264–280). Hershey, PA: IGI Global. doi:10.4018/978-1-4666-2160-2.ch016

Grassetti, M., & Brookby, S. (2013). Using the iPad to develop preservice teachers' understanding of the common core state standards for mathematical practice. In D. Polly (Ed.), *Common core mathematics standards and implementing digital technologies* (pp. 370–386). Hershey, PA: IGI Global. doi:10.4018/978-1-4666-4086-3.ch025

Gray, P. J. (2012). CDIO Standards and quality assurance: From application to accreditation. *International Journal of Quality Assurance in Engineering and Technology Education*, *2*(2), 1–8. doi:10.4018/ijqaete.2012040101

Graz, J., & Hauert, C. (2011). The INTERNORM project: Bridging two worlds of expert- and lay-knowledge in standardization. *International Journal of IT Standards and Standardization Research*, *9*(1), 52–62. doi:10.4018/jitsr.2011010103

Graz, J., & Hauert, C. (2013). The INTERNORM project: Bridging two worlds of expert- and lay-knowledge in standardization. In K. Jakobs (Ed.), *Innovations in organizational IT specification and standards development* (pp. 154–164). Hershey, PA: IGI Global. doi:10.4018/978-1-4666-2160-2.ch009

Grobler, M. (2012). The need for digital evidence standardisation. *International Journal of Digital Crime and Forensics*, *4*(2), 1–12. doi:10.4018/jdcf.2012040101

Grobler, M. (2013). The need for digital evidence standardisation. In C. Li (Ed.), *Emerging digital forensics applications for crime detection, prevention, and security* (pp. 234–245). Hershey, PA: IGI Global. doi:10.4018/978-1-4666-4006-1.ch016

Guest, C. L., & Guest, J. M. (2011). Legal issues in the use of technology in higher education: Copyright and privacy in the academy. In D. Surry, R. Gray Jr, & J. Stefurak (Eds.), *Technology integration in higher education: Social and organizational aspects* (pp. 72–85). Hershey, PA: IGI Global. doi:10.4018/978-1-60960-147-8.ch006

Gupta, A., Gantz, D. A., Sreecharana, D., & Kreyling, J. (2012). The interplay of offshoring of professional services, law, intellectual property, and international organizations. *International Journal of Strategic Information Technology and Applications*, *3*(2), 47–71. doi:10.4018/jsita.2012040104

Hai-Jew, S. (2011). Staying legal and ethical in global e-learning course and training developments: An exploration. In V. Wang (Ed.), *Encyclopedia of information communication technologies and adult education integration* (pp. 958–970). Hershey, PA: IGI Global. doi:10.4018/978-1-61692-906-0.ch058

Halder, D., & Jaishankar, K. (2012). Cyber space regulations for protecting women in UK. In *Cyber crime and the victimization of women: Laws, rights and regulations* (pp. 95–104). Hershey, PA: IGI Global. doi:10.4018/978-1-60960-830-9.ch007

Han, M., & Cho, C. (2013). XML in library cataloging workflows: Working with diverse sources and metadata standards. In J. Tramullas & P. Garrido (Eds.), *Library automation and OPAC 2.0: Information access and services in the 2.0 landscape* (pp. 59–72). Hershey, PA: IGI Global. doi:10.4018/978-1-4666-1912-8.ch003

Hanseth, O., & Nielsen, P. (2013). Infrastructural innovation: Flexibility, generativity and the mobile internet. *International Journal of IT Standards and Standardization Research*, *11*(1), 27–45. doi:10.4018/jitsr.2013010102

Hartong, M., & Wijesekera, D. (2012). U.S. regulatory requirements for positive train control systems. In F. Flammini (Ed.), *Railway safety, reliability, and security: Technologies and systems engineering* (pp. 1–21). Hershey, PA: IGI Global. doi:10.4018/978-1-4666-1643-1.ch001

Hasan, H. (2011). Formal and emergent standards in KM. In D. Schwartz & D. Te'eni (Eds.), *Encyclopedia of knowledge management* (2nd ed.; pp. 331–342). Hershey, PA: IGI Global. doi:10.4018/978-1-59904-931-1.ch032

Hatzimihail, N. (2011). Copyright infringement of digital libraries and private international law: Jurisdiction issues. In I. Iglezakis, T. Synodinou, & S. Kapidakis (Eds.), *E-publishing and digital libraries: Legal and organizational issues* (pp. 447–460). Hershey, PA: IGI Global. doi:10.4018/978-1-60960-031-0.ch021

**Related References**

Hauert, C. (2013). Where are you? Consumers' associations in standardization: A case study on Switzerland. In K. Jakobs (Ed.), *Innovations in organizational IT specification and standards development* (pp. 139–153). Hershey, PA: IGI Global. doi:10.4018/978-1-4666-2160-2.ch008

Hawks, V. D., & Ekstrom, J. J. (2011). Balancing policies, principles, and philosophy in information assurance. In M. Dark (Ed.), *Information assurance and security ethics in complex systems: Interdisciplinary perspectives* (pp. 32–54). Hershey, PA: IGI Global. doi:10.4018/978-1-61692-245-0.ch003

Henningsson, S. (2012). International e-customs standardization from the perspective of a global company. *International Journal of IT Standards and Standardization Research, 10*(2), 45–58. doi:10.4018/jitsr.2012070104

Hensberry, K. K., Paul, A. J., Moore, E. B., Podolefsky, N. S., & Perkins, K. K. (2013). PhET interactive simulations: New tools to achieve common core mathematics standards. In D. Polly (Ed.), *Common core mathematics standards and implementing digital technologies* (pp. 147–167). Hershey, PA: IGI Global. doi:10.4018/978-1-4666-4086-3.ch010

Heravi, B. R., & Lycett, M. (2012). Semantically enriched e-business standards development: The case of ebXML business process specification schema. In E. Kajan, F. Dorloff, & I. Bedini (Eds.), *Handbook of research on e-business standards and protocols: Documents, data and advanced web technologies* (pp. 655–675). Hershey, PA: IGI Global. doi:10.4018/978-1-4666-0146-8.ch030

Higuera, J., & Polo, J. (2012). Interoperability in wireless sensor networks based on IEEE 1451 standard. In N. Zaman, K. Ragab, & A. Abdullah (Eds.), *Wireless sensor networks and energy efficiency: Protocols, routing and management* (pp. 47–69). Hershey, PA: IGI Global. doi:10.4018/978-1-4666-0101-7.ch004

Hill, D. S. (2012). An examination of standardized product identification and business benefit. In E. Kajan, F. Dorloff, & I. Bedini (Eds.), *Handbook of research on e-business standards and protocols: Documents, data and advanced web technologies* (pp. 387–411). Hershey, PA: IGI Global. doi:10.4018/978-1-4666-0146-8.ch018

Hill, D. S. (2013). An examination of standardized product identification and business benefit. In *Supply chain management: Concepts, methodologies, tools, and applications* (pp. 171–195). Hershey, PA: IGI Global. doi:10.4018/978-1-4666-2625-6.ch011

Holloway, K. (2012). Fair use, copyright, and academic integrity in an online academic environment. In V. Wang (Ed.), *Encyclopedia of e-leadership, counseling and training* (pp. 298–309). Hershey, PA: IGI Global. doi:10.4018/978-1-61350-068-2.ch022

Hoops, D. S. (2011). Legal issues in the virtual world and e-commerce. In B. Ciaramitaro (Ed.), *Virtual worlds and e-commerce: Technologies and applications for building customer relationships* (pp. 186–204). Hershey, PA: IGI Global. doi:10.4018/978-1-61692-808-7.ch010

Hoops, D. S. (2012). Lost in cyberspace: Navigating the legal issues of e-commerce. *Journal of Electronic Commerce in Organizations, 10*(1), 33–51. doi:10.4018/jeco.2012010103

Hopkinson, A. (2012). Establishing the digital library: Don't ignore the library standards and don't forget the training needed. In A. Tella & A. Issa (Eds.), *Library and information science in developing countries: Contemporary issues* (pp. 195–204). Hershey, PA: IGI Global. doi:10.4018/978-1-61350-335-5.ch014

Hua, G. B. (2013). The construction industry and standardization of information. In *Implementing IT business strategy in the construction industry* (pp. 47–66). Hershey, PA: IGI Global. doi:10.4018/978-1-4666-4185-3.ch003

Huang, C., & Lin, H. (2011). Patent infringement risk analysis using rough set theory. In Q. Zhang, R. Segall, & M. Cao (Eds.), *Visual analytics and interactive technologies: Data, text and web mining applications* (pp. 123–150). Hershey, PA: IGI Global. doi:10.4018/978-1-60960-102-7.ch008

Huang, C., Tseng, T. B., & Lin, H. (2013). Patent infringement risk analysis using rough set theory. In *Digital rights management: Concepts, methodologies, tools, and applications* (pp. 1225–1251). Hershey, PA: IGI Global. doi:10.4018/978-1-4666-2136-7.ch060

Iyamu, T. (2013). The impact of organisational politics on the implementation of IT strategy: South African case in context. In J. Abdelnour-Nocera (Ed.), *Knowledge and technological development effects on organizational and social structures* (pp. 167–193). Hershey, PA: IGI Global. doi:10.4018/978-1-4666-2151-0.ch011

Jacinto, K., Neto, F. M., Leite, C. R., & Jacinto, K. (2014). Accessibility in u-learning: Standards, legislation, and future visions. In F. Neto (Ed.), *Technology platform innovations and forthcoming trends in ubiquitous learning* (pp. 215–236). Hershey, PA: IGI Global. doi:10.4018/978-1-4666-4542-4.ch012

**Related References**

Jakobs, K., Wagner, T., & Reimers, K. (2011). Standardising the internet of things: What the experts think. *International Journal of IT Standards and Standardization Research*, *9*(1), 63–67. doi:10.4018/jitsr.2011010104

Juzoji, H. (2012). Legal bases for medical supervision via mobile telecommunications in Japan. *International Journal of E-Health and Medical Communications*, *3*(1), 33–45. doi:10.4018/jehmc.2012010103

Kallinikou, D., Papadopoulos, M., Kaponi, A., & Strakantouna, V. (2011). Intellectual property issues for digital libraries at the intersection of law, technology, and the public interest. In I. Iglezakis, T. Synodinou, & S. Kapidakis (Eds.), *E-publishing and digital libraries: Legal and organizational issues* (pp. 294–341). Hershey, PA: IGI Global. doi:10.4018/978-1-60960-031-0.ch015

Kallinikou, D., Papadopoulos, M., Kaponi, A., & Strakantouna, V. (2013). Intellectual property issues for digital libraries at the intersection of law, technology, and the public interest. In *Digital rights management: Concepts, methodologies, tools, and applications* (pp. 1043–1090). Hershey, PA: IGI Global. doi:10.4018/978-1-4666-2136-7.ch052

Kaupins, G. (2012). Laws associated with mobile computing in the cloud. *International Journal of Wireless Networks and Broadband Technologies*, *2*(3), 1–9. doi:10.4018/ijwnbt.2012070101

Kaur, P., & Singh, H. (2013). Component certification process and standards. In H. Singh & K. Kaur (Eds.), *Designing, engineering, and analyzing reliable and efficient software* (pp. 22–39). Hershey, PA: IGI Global. doi:10.4018/978-1-4666-2958-5.ch002

Kayem, A. V. (2013). Security in service oriented architectures: Standards and challenges. In *Digital rights management: Concepts, methodologies, tools, and applications* (pp. 50–73). Hershey, PA: IGI Global. doi:10.4018/978-1-4666-2136-7.ch004

Kemp, M. L., Robb, S., & Deans, P. C. (2013). The legal implications of cloud computing. In A. Bento & A. Aggarwal (Eds.), *Cloud computing service and deployment models: Layers and management* (pp. 257–272). Hershey, PA: IGI Global. doi:10.4018/978-1-4666-2187-9.ch014

Khansa, L., & Liginlal, D. (2012). Regulatory influence and the imperative of innovation in identity and access management. *Information Resources Management Journal*, *25*(3), 78–97. doi:10.4018/irmj.2012070104

Kim, E. (2012). Government policies to promote production and consumption of renewable electricity in the US. In M. Tortora (Ed.), *Sustainable systems and energy management at the regional level: Comparative approaches* (pp. 1–18). Hershey, PA: IGI Global. doi:10.4018/978-1-61350-344-7.ch001

Kinsell, C. (2014). Technology and disability laws, regulations, and rights. In B. DaCosta & S. Seok (Eds.), *Assistive technology research, practice, and theory* (pp. 75–87). Hershey, PA: IGI Global. doi:10.4018/978-1-4666-5015-2.ch006

Kitsiou, S. (2010). Overview and analysis of electronic health record standards. In J. Rodrigues (Ed.), *Health information systems: Concepts, methodologies, tools, and applications* (pp. 374–392). Hershey, PA: IGI Global. doi:10.4018/978-1-60566-988-5.ch025

Kloss, J. H., & Schickel, P. (2011). X3D: A secure ISO standard for virtual worlds. In A. Rea (Ed.), *Security in virtual worlds, 3D webs, and immersive environments: Models for development, interaction, and management* (pp. 208–220). Hershey, PA: IGI Global. doi:10.4018/978-1-61520-891-3.ch010

Kotsonis, E., & Eliakis, S. (2011). Information security standards for health information systems: The implementer's approach. In A. Chryssanthou, I. Apostolakis, & I. Varlamis (Eds.), *Certification and security in health-related web applications: Concepts and solutions* (pp. 113–145). Hershey, PA: IGI Global. doi:10.4018/978-1-61692-895-7.ch006

Kotsonis, E., & Eliakis, S. (2013). Information security standards for health information systems: The implementer's approach. In *User-driven healthcare: Concepts, methodologies, tools, and applications* (pp. 225–257). Hershey, PA: IGI Global. doi:10.4018/978-1-4666-2770-3.ch013

Koumaras, H., & Kourtis, M. (2013). A survey on video coding principles and standards. In R. Farrugia & C. Debono (Eds.), *Multimedia networking and coding* (pp. 1–27). Hershey, PA: IGI Global. doi:10.4018/978-1-4666-2660-7.ch001

Krupinski, E. A., Antoniotti, N., & Burdick, A. (2011). Standards and guidelines development in the american telemedicine association. In A. Moumtzoglou & A. Kastania (Eds.), *E-health systems quality and reliability: Models and standards* (pp. 244–252). Hershey, PA: IGI Global. doi:10.4018/978-1-61692-843-8.ch018

### Related References

Kuanpoth, J. (2011). Biotechnological patents and morality: A critical view from a developing country. In S. Hongladarom (Ed.), *Genomics and bioethics: Interdisciplinary perspectives, technologies and advancements* (pp. 141–151). Hershey, PA: IGI Global. doi:10.4018/978-1-61692-883-4.ch010

Kuanpoth, J. (2013). Biotechnological patents and morality: A critical view from a developing country. In *Digital rights management: Concepts, methodologies, tools, and applications* (pp. 1417–1427). Hershey, PA: IGI Global. doi:10.4018/978-1-4666-2136-7.ch071

Kulmala, R., & Kettunen, J. (2012). Intellectual property protection and process modeling in small knowledge intensive enterprises. In *Organizational learning and knowledge: Concepts, methodologies, tools and applications* (pp. 2963–2980). Hershey, PA: IGI Global. doi:10.4018/978-1-60960-783-8.ch809

Kulmala, R., & Kettunen, J. (2013). Intellectual property protection in small knowledge intensive enterprises. *International Journal of Cyber Warfare & Terrorism*, *3*(1), 29–45. doi:10.4018/ijcwt.2013010103

Küster, M. W. (2012). Standards for achieving interoperability of egovernment in Europe. In E. Kajan, F. Dorloff, & I. Bedini (Eds.), *Handbook of research on e-business standards and protocols: Documents, data and advanced web technologies* (pp. 249–268). Hershey, PA: IGI Global. doi:10.4018/978-1-4666-0146-8.ch012

Kyobe, M. (2011). Factors influencing SME compliance with government regulation on use of IT: The case of South Africa. In F. Tan (Ed.), *International enterprises and global information technologies: Advancing management practices* (pp. 85–116). Hershey, PA: IGI Global. doi:10.4018/978-1-60960-605-3.ch005

Lam, J. C., & Hills, P. (2011). Promoting technological environmental innovations: What is the role of environmental regulation? In Z. Luo (Ed.), *Green finance and sustainability: Environmentally-aware business models and technologies* (pp. 56–73). Hershey, PA: IGI Global. doi:10.4018/978-1-60960-531-5.ch003

Lam, J. C., & Hills, P. (2013). Promoting technological environmental innovations: The role of environmental regulation. In Z. Luo (Ed.), *Technological solutions for modern logistics and supply chain management* (pp. 230–247). Hershey, PA: IGI Global. doi:10.4018/978-1-4666-2773-4.ch015

Laporte, C., & Vargas, E. P. (2012). The development of international standards to facilitate process improvements for very small entities. In S. Fauzi, M. Nasir, N. Ramli, & S. Sahibuddin (Eds.), *Software process improvement and management: Approaches and tools for practical development* (pp. 34–61). Hershey, PA: IGI Global. doi:10.4018/978-1-61350-141-2.ch003

Laporte, C., & Vargas, E. P. (2014). The development of international standards to facilitate process improvements for very small entities. In *Software design and development: Concepts, methodologies, tools, and applications* (pp. 1335–1361). Hershey, PA: IGI Global. doi:10.4018/978-1-4666-4301-7.ch065

Lautman, R., & Curran, K. (2013). The problems of jurisdiction on the internet. In K. Curran (Ed.), *Pervasive and ubiquitous technology innovations for ambient intelligence environments* (pp. 164–170). Hershey, PA: IGI Global. doi:10.4018/978-1-4666-2041-4.ch016

Layne-Farrar, A. (2011). Innovative or indefensible? An empirical assessment of patenting within standard setting. *International Journal of IT Standards and Standardization Research, 9*(2), 1–18. doi:10.4018/jitsr.2011070101

Layne-Farrar, A. (2013). Innovative or indefensible? An empirical assessment of patenting within standard setting. In K. Jakobs (Ed.), *Innovations in organizational IT specification and standards development* (pp. 1–18). Hershey, PA: IGI Global. doi:10.4018/978-1-4666-2160-2.ch001

Layne-Farrar, A., & Padilla, A. J. (2011). Assessing the link between standards and patents. *International Journal of IT Standards and Standardization Research, 9*(2), 19–49. doi:10.4018/jitsr.2011070102

Layne-Farrar, A., & Padilla, A. J. (2013). Assessing the link between standards and patents. In K. Jakobs (Ed.), *Innovations in organizational IT specification and standards development* (pp. 19–51). Hershey, PA: IGI Global. doi:10.4018/978-1-4666-2160-2.ch002

Lee, H., & Huh, J. C. (2012). Koreas strategies for ICT standards internationalisation: A comparison with Chinas. *International Journal of IT Standards and Standardization Research, 10*(2), 1–13. doi:10.4018/jitsr.2012070101

Li, Y., & Wei, C. (2011). Digital image authentication: A review. *International Journal of Digital Library Systems, 2*(2), 55–78. doi:10.4018/jdls.2011040104

**Related References**

Li, Y., Xiao, X., Feng, X., & Yan, H. (2012). Adaptation and localization: Metadata research and development for Chinese digital resources. *International Journal of Digital Library Systems*, *3*(1), 1–21. doi:10.4018/jdls.2012010101

Lim, W., & Kim, D. (2013). Do technologies support the implementation of the common core state standards in mathematics of high school probability and statistics? In D. Polly (Ed.), *Common core mathematics standards and implementing digital technologies* (pp. 168–183). Hershey, PA: IGI Global. doi:10.4018/978-1-4666-4086-3.ch011

Linton, J., & Stegall, D. (2013). Common core standards for mathematical practice and TPACK: An integrated approach to instruction. In D. Polly (Ed.), *Common core mathematics standards and implementing digital technologies* (pp. 234–249). Hershey, PA: IGI Global. doi:10.4018/978-1-4666-4086-3.ch016

Liotta, A., & Liotta, A. (2011). Privacy in pervasive systems: Legal framework and regulatory challenges. In A. Malatras (Ed.), *Pervasive computing and communications design and deployment: Technologies, trends and applications* (pp. 263–277). Hershey, PA: IGI Global. doi:10.4018/978-1-60960-611-4.ch012

Lissoni, F. (2013). Academic patenting in Europe: Recent research and new perspectives. In I. Yetkiner, M. Pamukcu, & E. Erdil (Eds.), *Industrial dynamics, innovation policy, and economic growth through technological advancements* (pp. 75–91). Hershey, PA: IGI Global. doi:10.4018/978-1-4666-1978-4.ch005

Litaay, T., Prananingrum, D. H., & Krisanto, Y. A. (2011). Indonesian legal perspectives on biotechnology and intellectual property rights. In S. Hongladarom (Ed.), *Genomics and bioethics: Interdisciplinary perspectives, technologies and advancements* (pp. 171–183). Hershey, PA: IGI Global. doi:10.4018/978-1-61692-883-4.ch012

Litaay, T., Prananingrum, D. H., & Krisanto, Y. A. (2013). Indonesian legal perspectives on biotechnology and intellectual property rights. In *Digital rights management: Concepts, methodologies, tools, and applications* (pp. 834–845). Hershey, PA: IGI Global. doi:10.4018/978-1-4666-2136-7.ch039

Losavio, M., Pastukhov, P., & Polyakova, S. (2014). Regulatory aspects of cloud computing in business environments. In S. Srinivasan (Ed.), *Security, trust, and regulatory aspects of cloud computing in business environments* (pp. 156–169). Hershey, PA: IGI Global. doi:10.4018/978-1-4666-5788-5.ch009

Lu, B., Tsou, B. K., Jiang, T., Zhu, J., & Kwong, O. Y. (2011). Mining parallel knowledge from comparable patents. In W. Wong, W. Liu, & M. Bennamoun (Eds.), *Ontology learning and knowledge discovery using the web: Challenges and recent advances* (pp. 247–271). Hershey, PA: IGI Global. doi:10.4018/978-1-60960-625-1.ch013

Lucas-Schloetter, A. (2011). Digital libraries and copyright issues: Digitization of contents and the economic rights of the authors. In I. Iglezakis, T. Synodinou, & S. Kapidakis (Eds.), *E-publishing and digital libraries: Legal and organizational issues* (pp. 159–179). Hershey, PA: IGI Global. doi:10.4018/978-1-60960-031-0.ch009

Lyytinen, K., Keil, T., & Fomin, V. (2010). A framework to build process theories of anticipatory information and communication technology (ICT) standardizing. In K. Jakobs (Ed.), *New applications in IT standards: Developments and progress* (pp. 147–186). Hershey, PA: IGI Global. doi:10.4018/978-1-60566-946-5.ch008

Macedo, M., & Isaías, P. (2013). Standards related to interoperability in EHR & HS. In M. Sicilia & P. Balazote (Eds.), *Interoperability in healthcare information systems: Standards, management, and technology* (pp. 19–44). Hershey, PA: IGI Global. doi:10.4018/978-1-4666-3000-0.ch002

Madden, P. (2011). Greater accountability, less red tape: The Australian standard business reporting experience. *International Journal of E-Business Research*, 7(2), 1–10. doi:10.4018/jebr.2011040101

Maravilhas, S. (2014). Quality improves the value of patent information to promote innovation. In G. Jamil, A. Malheiro, & F. Ribeiro (Eds.), *Rethinking the conceptual base for new practical applications in information value and quality* (pp. 61–82). Hershey, PA: IGI Global. doi:10.4018/978-1-4666-4562-2.ch004

Marshall, S. (2011). E-learning standards: Beyond technical standards to guides for professional practice. In F. Lazarinis, S. Green, & E. Pearson (Eds.), *Handbook of research on e-learning standards and interoperability: Frameworks and issues* (pp. 170–192). Hershey, PA: IGI Global. doi:10.4018/978-1-61692-789-9.ch008

Martino, L., & Bertino, E. (2012). Security for web services: Standards and research issues. In L. Jie-Zhang (Ed.), *Innovations, standards and practices of web services: Emerging research topics* (pp. 336–362). Hershey, PA: IGI Global. doi:10.4018/978-1-61350-104-7.ch015

***Related References***

McCarthy, V., & Hulsart, R. (2012). Management education for integrity: Raising ethical standards in online management classes. In C. Wankel & A. Stachowicz-Stanusch (Eds.), *Handbook of research on teaching ethics in business and management education* (pp. 413–425). Hershey, PA: IGI Global. doi:10.4018/978-1-61350-510-6.ch024

McGrath, T. (2012). The reality of using standards for electronic business document formats. In E. Kajan, F. Dorloff, & I. Bedini (Eds.), *Handbook of research on e-business standards and protocols: Documents, data and advanced web technologies* (pp. 21–32). Hershey, PA: IGI Global. doi:10.4018/978-1-4666-0146-8.ch002

Medlin, B. D., & Chen, C. C. (2012). A global perspective of laws and regulations dealing with information security and privacy. In *Cyber crime: Concepts, methodologies, tools and applications* (pp. 1349–1363). Hershey, PA: IGI Global. doi:10.4018/978-1-61350-323-2.ch609

Mehrfard, H., & Hamou-Lhadj, A. (2011). The impact of regulatory compliance on agile software processes with a focus on the FDA guidelines for medical device software. *International Journal of Information System Modeling and Design*, *2*(2), 67–81. doi:10.4018/jismd.2011040104

Mehrfard, H., & Hamou-Lhadj, A. (2013). The impact of regulatory compliance on agile software processes with a focus on the FDA guidelines for medical device software. In J. Krogstie (Ed.), *Frameworks for developing efficient information systems: Models, theory, and practice* (pp. 298–314). Hershey, PA: IGI Global. doi:10.4018/978-1-4666-4161-7.ch013

Mendoza, R. A., & Ravichandran, T. (2011). An exploratory analysis of the relationship between organizational and institutional factors shaping the assimilation of vertical standards. *International Journal of IT Standards and Standardization Research*, *9*(1), 24–51. doi:10.4018/jitsr.2011010102

Mendoza, R. A., & Ravichandran, T. (2012). An empirical evaluation of the assimilation of industry-specific data standards using firm-level and community-level constructs. In M. Tavana (Ed.), *Enterprise information systems and advancing business solutions: Emerging models* (pp. 287–312). Hershey, PA: IGI Global. doi:10.4018/978-1-4666-1761-2.ch017

Mendoza, R. A., & Ravichandran, T. (2012). Drivers of organizational participation in XML-based industry standardization efforts. In M. Tavana (Ed.), *Enterprise information systems and advancing business solutions: Emerging models* (pp. 268–286). Hershey, PA: IGI Global. doi:10.4018/978-1-4666-1761-2.ch016

Mendoza, R. A., & Ravichandran, T. (2013). An exploratory analysis of the relationship between organizational and institutional factors shaping the assimilation of vertical standards. In K. Jakobs (Ed.), *Innovations in organizational IT specification and standards development* (pp. 193–221). Hershey, PA: IGI Global. doi:10.4018/978-1-4666-2160-2.ch012

Mense, E. G., Fulwiler, J. H., Richardson, M. D., & Lane, K. E. (2011). Standardization, hybridization, or individualization: Marketing IT to a diverse clientele. In U. Demiray & S. Sever (Eds.), *Marketing online education programs: Frameworks for promotion and communication* (pp. 291–299). Hershey, PA: IGI Global. doi:10.4018/978-1-60960-074-7.ch019

Metaxa, E., Sarigiannidis, M., & Folinas, D. (2012). Legal issues of the French law on creation and internet (Hadopi 1 and 2). *International Journal of Technoethics*, *3*(3), 21–36. doi:10.4018/jte.2012070102

Meyer, N. (2012). Standardization as governance without government: A critical reassessment of the digital video broadcasting projects success story. *International Journal of IT Standards and Standardization Research*, *10*(2), 14–28. doi:10.4018/jitsr.2012070102

Miguel da Silva, F., Neto, F. M., Burlamaqui, A. M., Pinto, J. P., Fernandes, C. E., & Castro de Souza, R. (2014). T-SCORM: An extension of the SCORM standard to support the project of educational contents for t-learning. In F. Neto (Ed.), *Technology platform innovations and forthcoming trends in ubiquitous learning* (pp. 94–119). Hershey, PA: IGI Global. doi:10.4018/978-1-4666-4542-4.ch006

Moon, A. (2014). Copyright and licensing essentials for librarians and copyright owners in the digital age. In N. Patra, B. Kumar, & A. Pani (Eds.), *Progressive trends in electronic resource management in libraries* (pp. 106–117). Hershey, PA: IGI Global. doi:10.4018/978-1-4666-4761-9.ch006

Moralis, A., Pouli, V., Grammatikou, M., Kalogeras, D., & Maglaris, V. (2012). Security standards and issues for grid computing. In *Grid and cloud computing: Concepts, methodologies, tools and applications* (pp. 1656–1671). Hershey, PA: IGI Global. doi:10.4018/978-1-4666-0879-5.ch708

Moreno, L., Iglesias, A., Calvo, R., Delgado, S., & Zaragoza, L. (2012). Disability standards and guidelines for learning management systems: Evaluating accessibility. In R. Babo & A. Azevedo (Eds.), *Higher education institutions and learning management systems: Adoption and standardization* (pp. 199–218). Hershey, PA: IGI Global. doi:10.4018/978-1-60960-884-2.ch010

**Related References**

Moro, N. (2013). Digital rights management and corporate hegemony: Avenues for reform. In H. Rahman & I. Ramos (Eds.), *Ethical data mining applications for socio-economic development* (pp. 281–299). Hershey, PA: IGI Global. doi:10.4018/978-1-4666-4078-8.ch013

Mula, D., & Lobina, M. L. (2012). Legal protection of the web page. In H. Sasaki (Ed.), *Information technology for intellectual property protection: Interdisciplinary advancements* (pp. 213–236). Hershey, PA: IGI Global. doi:10.4018/978-1-61350-135-1.ch008

Mula, D., & Lobina, M. L. (2013). Legal protection of the web page. In *Digital rights management: Concepts, methodologies, tools, and applications* (pp. 1–18). Hershey, PA: IGI Global. doi:10.4018/978-1-4666-2136-7.ch001

Mulcahy, D. (2011). Performativity in practice: An actor-network account of professional teaching standards. *International Journal of Actor-Network Theory and Technological Innovation, 3*(2), 1–16. doi:10.4018/jantti.2011040101

Mulcahy, D. (2013). Performativity in practice: An actor-network account of professional teaching standards. In A. Tatnall (Ed.), *Social and professional applications of actor-network theory for technology development* (pp. 1–16). Hershey, PA: IGI Global. doi:10.4018/978-1-4666-2166-4.ch001

Mustaffa, M. T. (2012). Multi-standard multi-band reconfigurable LNA. In A. Marzuki, A. Rahim, & M. Loulou (Eds.), *Advances in monolithic microwave integrated circuits for wireless systems: Modeling and design technologies* (pp. 1–23). Hershey, PA: IGI Global. doi:10.4018/978-1-60566-886-4.ch001

Nabi, S. I., Al-Ghmlas, G. S., & Alghathbar, K. (2012). Enterprise information security policies, standards, and procedures: A survey of available standards and guidelines. In M. Gupta, J. Walp, & R. Sharman (Eds.), *Strategic and practical approaches for information security governance: Technologies and applied solutions* (pp. 67–89). Hershey, PA: IGI Global. doi:10.4018/978-1-4666-0197-0.ch005

Nabi, S. I., Al-Ghmlas, G. S., & Alghathbar, K. (2014). Enterprise information security policies, standards, and procedures: A survey of available standards and guidelines. In *Crisis management: Concepts, methodologies, tools and applications* (pp. 750–773). Hershey, PA: IGI Global. doi:10.4018/978-1-4666-4707-7.ch036

Naixiao, Z., & Chunhua, H. (2012). Research on open innovation in China: Focus on intellectual property rights and their operation in Chinese enterprises. *International Journal of Asian Business and Information Management, 3*(1), 65–71. doi:10.4018/jabim.2012010106

Naixiao, Z., & Chunhua, H. (2013). Research on open innovation in China: Focus on intellectual property rights and their operation in Chinese enterprises. In *Digital rights management: Concepts, methodologies, tools, and applications* (pp. 714–720). Hershey, PA: IGI Global. doi:10.4018/978-1-4666-2136-7.ch031

Ndjetcheu, L. (2013). Social responsibility and legal financial communication in African companies in the south of the Sahara: Glance from the OHADA accounting law viewpoint. *International Journal of Innovation in the Digital Economy*, *4*(4), 1–17. doi:10.4018/ijide.2013100101

Ng, W. L. (2013). Improving long-term financial risk forecasts using high-frequency data and scaling laws. In B. Alexandrova-Kabadjova, S. Martinez-Jaramillo, A. Garcia-Almanza, & E. Tsang (Eds.), *Simulation in computational finance and economics: Tools and emerging applications* (pp. 255–278). Hershey, PA: IGI Global. doi:10.4018/978-1-4666-2011-7.ch013

Noury, N., Bourquard, K., Bergognon, D., & Schroeder, J. (2013). Regulations initiatives in France for the interoperability of communicating medical devices. *International Journal of E-Health and Medical Communications*, *4*(2), 50–64. doi:10.4018/jehmc.2013040104

Null, E. (2013). Legal and political barriers to municipal networks in the United States. In A. Abdelaal (Ed.), *Social and economic effects of community wireless networks and infrastructures* (pp. 27–56). Hershey, PA: IGI Global. doi:10.4018/978-1-4666-2997-4.ch003

OConnor, R. V., & Laporte, C. Y. (2014). An innovative approach to the development of an international software process lifecycle standard for very small entities. *International Journal of Information Technologies and Systems Approach*, *7*(1), 1–22. doi:10.4018/ijitsa.2014010101

Onat, I., & Miri, A. (2013). RFID standards. In A. Miri (Ed.), *Advanced security and privacy for RFID technologies* (pp. 14–22). Hershey, PA: IGI Global. doi:10.4018/978-1-4666-3685-9.ch002

Orton, I., Alva, A., & Endicott-Popovsky, B. (2013). Legal process and requirements for cloud forensic investigations. In K. Ruan (Ed.), *Cybercrime and cloud forensics: Applications for investigation processes* (pp. 186–229). Hershey, PA: IGI Global. doi:10.4018/978-1-4666-2662-1.ch008

***Related References***

Ortt, J. R., & Egyedi, T. M. (2014). The effect of pre-existing standards and regulations on the development and diffusion of radically new innovations. *International Journal of IT Standards and Standardization Research, 12*(1), 17–37. doi:10.4018/ijitsr.2014010102

Ozturk, Y., & Sharma, J. (2011). mVITAL: A standards compliant vital sign monitor. In C. Röcker, & M. Ziefle (Eds.), Smart healthcare applications and services: Developments and practices (pp. 174-196). Hershey, PA: IGI Global. doi:10.4018/978-1-60960-180-5.ch008

Ozturk, Y., & Sharma, J. (2013). mVITAL: A standards compliant vital sign monitor. In IT policy and ethics: Concepts, methodologies, tools, and applications (pp. 515-538). Hershey, PA: IGI Global. doi:10.4018/978-1-4666-2919-6.ch024

Parsons, T. D. (2011). Affect-sensitive virtual standardized patient interface system. In D. Surry, R. Gray Jr, & J. Stefurak (Eds.), *Technology integration in higher education: Social and organizational aspects* (pp. 201–221). Hershey, PA: IGI Global. doi:10.4018/978-1-60960-147-8.ch015

Parveen, S., & Pater, C. (2012). Utilizing innovative video chat technology to meet national standards: A Case study on a STARTALK Hindi language program. *International Journal of Virtual and Personal Learning Environments, 3*(3), 1–20. doi:10.4018/jvple.2012070101

Pawlowski, J. M., & Kozlov, D. (2013). Analysis and validation of learning technology models, standards and specifications: The reference model analysis grid (RMAG). In K. Jakobs (Ed.), *Innovations in organizational IT specification and standards development* (pp. 223–240). Hershey, PA: IGI Global. doi:10.4018/978-1-4666-2160-2.ch013

Pina, P. (2011). The private copy issue: Piracy, copyright and consumers' rights. In T. Strader (Ed.), *Digital product management, technology and practice: Interdisciplinary perspectives* (pp. 193–205). Hershey, PA: IGI Global. doi:10.4018/978-1-61692-877-3.ch011

Pina, P. (2013). Between Scylla and Charybdis: The balance between copyright, digital rights management and freedom of expression. In Digital rights management: Concepts, methodologies, tools, and applications (pp. 1355-1367). Hershey, PA: IGI Global. doi:10.4018/978-1-4666-2136-7.ch067

Pina, P. (2013). Computer games and intellectual property law: Derivative works, copyright and copyleft. In *Digital rights management: Concepts, methodologies, tools, and applications* (pp. 777–788). Hershey, PA: IGI Global. doi:10.4018/978-1-4666-2136-7.ch035

Pina, P. (2013). The private copy issue: Piracy, copyright and consumers' rights. In *Digital rights management: Concepts, methodologies, tools, and applications* (pp. 1546–1558). Hershey, PA: IGI Global. doi:10.4018/978-1-4666-2136-7.ch078

Piotrowski, M. (2011). QTI: A failed e-learning standard? In F. Lazarinis, S. Green, & E. Pearson (Eds.), *Handbook of research on e-learning standards and interoperability: Frameworks and issues* (pp. 59–82). Hershey, PA: IGI Global. doi:10.4018/978-1-61692-789-9.ch004

Ponte, D., & Camussone, P. F. (2013). Neither heroes nor chaos: The victory of VHS against Betamax. *International Journal of Actor-Network Theory and Technological Innovation*, 5(1), 40–54. doi:10.4018/jantti.2013010103

Pradhan, A. (2011). Pivotal role of the ISO 14001 standard in the carbon economy. *International Journal of Green Computing*, 2(1), 38–46. doi:10.4018/jgc.2011010104

Pradhan, A. (2011). Standards and legislation for the carbon economy. In B. Unhelkar (Ed.), *Handbook of research on green ICT: Technology, business and social perspectives* (pp. 592–606). Hershey, PA: IGI Global. doi:10.4018/978-1-61692-834-6.ch043

Pradhan, A. (2013). Pivotal role of the ISO 14001 standard in the carbon economy. In K. Ganesh & S. Anbuudayasankar (Eds.), *International and interdisciplinary studies in green computing* (pp. 38–46). Hershey, PA: IGI Global. doi:10.4018/978-1-4666-2646-1.ch004

Prentzas, J., & Hatzilygeroudis, I. (2011). Techniques, technologies and patents related to intelligent educational systems. In G. Magoulas (Ed.), *E-infrastructures and technologies for lifelong learning: Next generation environments* (pp. 1–28). Hershey, PA: IGI Global. doi:10.4018/978-1-61520-983-5.ch001

Ramos, I., & Fernandes, J. (2011). Web-based intellectual property marketplace: A survey of current practices. *International Journal of Information Communication Technologies and Human Development*, 3(3), 58–68. doi:10.4018/jicthd.2011070105

**Related References**

Ramos, I., & Fernandes, J. (2013). Web-based intellectual property marketplace: A survey of current practices. In S. Chhabra (Ed.), *ICT influences on human development, interaction, and collaboration* (pp. 203–213). Hershey, PA: IGI Global. doi:10.4018/978-1-4666-1957-9.ch012

Rashmi, R. (2011). Biopharma drugs innovation in India and foreign investment and technology transfer in the changed patent regime. In P. Ordóñez de Pablos, W. Lee, & J. Zhao (Eds.), *Regional innovation systems and sustainable development: Emerging technologies* (pp. 210–225). Hershey, PA: IGI Global. doi:10.4018/978-1-61692-846-9.ch016

Rashmi, R. (2011). Optimal policy for biopharmaceutical drugs innovation and access in India. In P. Ordóñez de Pablos, W. Lee, & J. Zhao (Eds.), *Regional innovation systems and sustainable development: Emerging technologies* (pp. 74–114). Hershey, PA: IGI Global. doi:10.4018/978-1-61692-846-9.ch007

Rashmi, R. (2013). Biopharma drugs innovation in India and foreign investment and technology transfer in the changed patent regime. In *Digital rights management: Concepts, methodologies, tools, and applications* (pp. 846–859). Hershey, PA: IGI Global. doi:10.4018/978-1-4666-2136-7.ch040

Reed, C. N. (2011). The open geospatial consortium and web services standards. In P. Zhao & L. Di (Eds.), *Geospatial web services: Advances in information interoperability* (pp. 1–16). Hershey, PA: IGI Global. doi:10.4018/978-1-60960-192-8.ch001

Rejas-Muslera, R., Davara, E., Abran, A., & Buglione, L. (2013). Intellectual property systems in software. *International Journal of Cyber Warfare & Terrorism, 3*(1), 1–14. doi:10.4018/ijcwt.2013010101

Rejas-Muslera, R. J., García-Tejedor, A. J., & Rodriguez, O. P. (2011). Open educational resources in e-learning: standards and environment. In F. Lazarinis, S. Green, & E. Pearson (Eds.), *Handbook of research on e-learning standards and interoperability: Frameworks and issues* (pp. 346–359). Hershey, PA: IGI Global. doi:10.4018/978-1-61692-789-9.ch017

Ries, N. M. (2011). Legal issues in health information and electronic health records. In *Clinical technologies: Concepts, methodologies, tools and applications* (pp. 1948–1961). Hershey, PA: IGI Global. doi:10.4018/978-1-60960-561-2.ch708

Riillo, C. A. (2013). Profiles and motivations of standardization players. *International Journal of IT Standards and Standardization Research, 11*(2), 17–33. doi:10.4018/jitsr.2013070102

Rodriguez, E., & Lolas, F. (2011). Social issues related to gene patenting in Latin America: A bioethical reflection. In S. Hongladarom (Ed.), *Genomics and bioethics: Interdisciplinary perspectives, technologies and advancements* (pp. 152–170). Hershey, PA: IGI Global. doi:10.4018/978-1-61692-883-4.ch011

Rutherford, M. (2013). Implementing common core state standards using digital curriculum. In D. Polly (Ed.), *Common core mathematics standards and implementing digital technologies* (pp. 38–44). Hershey, PA: IGI Global. doi:10.4018/978-1-4666-4086-3.ch003

Rutherford, M. (2014). Implementing common core state standards using digital curriculum. In *K-12 education: Concepts, methodologies, tools, and applications* (pp. 383–389). Hershey, PA: IGI Global. doi:10.4018/978-1-4666-4502-8.ch022

Ryan, G., & Shinnick, E. (2011). Knowledge and intellectual property rights: An economics perspective. In D. Schwartz & D. Te'eni (Eds.), *Encyclopedia of knowledge management* (2nd ed.; pp. 489–496). Hershey, PA: IGI Global. doi:10.4018/978-1-59904-931-1.ch047

Ryoo, J., & Choi, Y. (2011). A taxonomy of green information and communication protocols and standards. In B. Unhelkar (Ed.), *Handbook of research on green ICT: Technology, business and social perspectives* (pp. 364–376). Hershey, PA: IGI Global. doi:10.4018/978-1-61692-834-6.ch026

Saeed, K., Ziegler, G., & Yaqoob, M. K. (2013). Management practices in exploration and production industry. In S. Saeed, M. Khan, & R. Ahmad (Eds.), *Business strategies and approaches for effective engineering management* (pp. 151–187). Hershey, PA: IGI Global. doi:10.4018/978-1-4666-3658-3.ch010

Saiki, T. (2014). Intellectual property in mergers & acquisitions. In J. Wang (Ed.), *Encyclopedia of business analytics and optimization* (pp. 1275–1283). Hershey, PA: IGI Global. doi:10.4018/978-1-4666-5202-6.ch117

Santos, O., & Boticario, J. (2011). A general framework for inclusive lifelong learning in higher education institutions with adaptive web-based services that support standards. In G. Magoulas (Ed.), *E-infrastructures and technologies for lifelong learning: Next generation environments* (pp. 29–58). Hershey, PA: IGI Global. doi:10.4018/978-1-61520-983-5.ch002

### Related References

Santos, O., Boticario, J., Raffenne, E., Granado, J., Rodriguez-Ascaso, A., & Gutierrez y Restrepo, E. (2011). A standard-based framework to support personalisation, adaptation, and interoperability in inclusive learning scenarios. In F. Lazarinis, S. Green, & E. Pearson (Eds.), *Handbook of research on e-learning standards and interoperability: Frameworks and issues* (pp. 126–169). Hershey, PA: IGI Global. doi:10.4018/978-1-61692-789-9.ch007

Sarabdeen, J. (2012). Legal issues in e-healthcare systems. In M. Watfa (Ed.), *E-healthcare systems and wireless communications: Current and future challenges* (pp. 23–48). Hershey, PA: IGI Global. doi:10.4018/978-1-61350-123-8.ch002

Scheg, A. G. (2014). Common standards for online education found in accrediting organizations. In *Reforming teacher education for online pedagogy development* (pp. 50–76). Hershey, PA: IGI Global. doi:10.4018/978-1-4666-5055-8.ch003

Sclater, N. (2012). Legal and contractual issues of cloud computing for educational institutions. In L. Chao (Ed.), *Cloud computing for teaching and learning: Strategies for design and implementation* (pp. 186–199). Hershey, PA: IGI Global. doi:10.4018/978-1-4666-0957-0.ch013

Selwyn, L., & Eldridge, V. (2013). Governance and organizational structures. In *Public law librarianship: Objectives, challenges, and solutions* (pp. 41–71). Hershey, PA: IGI Global. doi:10.4018/978-1-4666-2184-8.ch003

Seo, D. (2012). The significance of government's role in technology standardization: Two cases in the wireless communications industry. In C. Reddick (Ed.), *Cases on public information management and e-government adoption* (pp. 219–231). Hershey, PA: IGI Global. doi:10.4018/978-1-4666-0981-5.ch009

Seo, D. (2013). Analysis of various structures of standards setting organizations (SSOs) that impact tension among members. *International Journal of IT Standards and Standardization Research*, *11*(2), 46–60. doi:10.4018/jitsr.2013070104

Seo, D. (2013). Background of standards strategy. In *Evolution and standardization of mobile communications technology* (pp. 1–17). Hershey, PA: IGI Global. doi:10.4018/978-1-4666-4074-0.ch001

Seo, D. (2013). Developing a theoretical model. In *Evolution and standardization of mobile communications technology* (pp. 18–42). Hershey, PA: IGI Global. doi:10.4018/978-1-4666-4074-0.ch002

Seo, D. (2013). The 1G (first generation) mobile communications technology standards. In *Evolution and standardization of mobile communications technology* (pp. 54–75). Hershey, PA: IGI Global. doi:10.4018/978-1-4666-4074-0.ch005

Seo, D. (2013). The 2G (second generation) mobile communications technology standards. In *Evolution and standardization of mobile communications technology* (pp. 76–114). Hershey, PA: IGI Global. doi:10.4018/978-1-4666-4074-0.ch006

Seo, D. (2013). The 3G (third generation) of mobile communications technology standards. In *Evolution and standardization of mobile communications technology* (pp. 115–161). Hershey, PA: IGI Global. doi:10.4018/978-1-4666-4074-0.ch007

Seo, D. (2013). The significance of government's role in technology standardization: Two cases in the wireless communications industry. In K. Jakobs (Ed.), *Innovations in organizational IT specification and standards development* (pp. 183–192). Hershey, PA: IGI Global. doi:10.4018/978-1-4666-2160-2.ch011

Seo, D., & Koek, J. W. (2012). Are Asian countries ready to lead a global ICT standardization? *International Journal of IT Standards and Standardization Research*, *10*(2), 29–44. doi:10.4018/jitsr.2012070103

Sharp, R. J., Ewald, J. A., & Kenward, R. (2013). Central information flows and decision-making requirements. In J. Papathanasiou, B. Manos, S. Arampatzis, & R. Kenward (Eds.), *Transactional environmental support system design: Global solutions* (pp. 7–32). Hershey, PA: IGI Global. doi:10.4018/978-1-4666-2824-3.ch002

Shen, X., Graham, I., Stewart, J., & Williams, R. (2013). Standards development as hybridization. *International Journal of IT Standards and Standardization Research*, *11*(2), 34–45. doi:10.4018/jitsr.2013070103

Sherman, M. (2013). Using technology to engage students with the standards for mathematical practice: The case of DGS. In D. Polly (Ed.), *Common core mathematics standards and implementing digital technologies* (pp. 78–101). Hershey, PA: IGI Global. doi:10.4018/978-1-4666-4086-3.ch006

Singh, J., & Kumar, V. (2013). Compliance and regulatory standards for cloud computing. In R. Khurana & R. Aggarwal (Eds.), *Interdisciplinary perspectives on business convergence, computing, and legality* (pp. 54–64). Hershey, PA: IGI Global. doi:10.4018/978-1-4666-4209-6.ch006

**Related References**

Singh, S., & Paliwal, M. (2014). Exploring a sense of intellectual property valuation for Indian SMEs. *International Journal of Asian Business and Information Management, 5*(1), 15–36. doi:10.4018/ijabim.2014010102

Singh, S., & Siddiqui, T. J. (2013). Robust image data hiding technique for copyright protection. *International Journal of Information Security and Privacy, 7*(2), 44–56. doi:10.4018/jisp.2013040103

Spies, M., & Tabet, S. (2012). Emerging standards and protocols for governance, risk, and compliance management. In E. Kajan, F. Dorloff, & I. Bedini (Eds.), *Handbook of research on e-business standards and protocols: Documents, data and advanced web technologies* (pp. 768–790). Hershey, PA: IGI Global. doi:10.4018/978-1-4666-0146-8.ch035

Spinello, R. A., & Tavani, H. T. (2008). Intellectual property rights: From theory to practical implementation. In H. Sasaki (Ed.), *Intellectual property protection for multimedia information technology* (pp. 25–69). Hershey, PA: IGI Global. doi:10.4018/978-1-59904-762-1.ch002

Spyrou, S., Bamidis, P., & Maglaveras, N. (2010). Health information standards: Towards integrated health information networks. In J. Rodrigues (Ed.), *Health information systems: Concepts, methodologies, tools, and applications* (pp. 2145–2159). Hershey, PA: IGI Global. doi:10.4018/978-1-60566-988-5.ch136

Stanfill, D. (2012). Standards-based educational technology professional development. In V. Wang (Ed.), *Encyclopedia of e-leadership, counseling and training* (pp. 819–834). Hershey, PA: IGI Global. doi:10.4018/978-1-61350-068-2.ch060

Steen, H. U. (2011). The battle within: An analysis of internal fragmentation in networked technologies based on a comparison of the DVB-H and T-DMB mobile digital multimedia broadcasting standards. *International Journal of IT Standards and Standardization Research, 9*(2), 50–71. doi:10.4018/jitsr.2011070103

Steen, H. U. (2013). The battle within: An analysis of internal fragmentation in networked technologies based on a comparison of the DVB-H and T-DMB mobile digital multimedia broadcasting standards. In K. Jakobs (Ed.), *Innovations in organizational IT specification and standards development* (pp. 91–114). Hershey, PA: IGI Global. doi:10.4018/978-1-4666-2160-2.ch005

Stoll, M., & Breu, R. (2012). Information security governance and standard based management systems. In M. Gupta, J. Walp, & R. Sharman (Eds.), *Strategic and practical approaches for information security governance: Technologies and applied solutions* (pp. 261–282). Hershey, PA: IGI Global. doi:10.4018/978-1-4666-0197-0.ch015

Suzuki, O. (2013). Search efforts, selective appropriation, and the usefulness of new knowledge: Evidence from a comparison across U.S. and non-U.S. patent applicants. *International Journal of Knowledge Management, 9*(1), 42-59. doi:10.4018/jkm.2013010103

Tajima, M. (2012). The role of technology standardization in RFID adoption: The pharmaceutical context. *International Journal of IT Standards and Standardization Research, 10*(1), 48–67. doi:10.4018/jitsr.2012010104

Talevi, A., Castro, E. A., & Bruno-Blanch, L. E. (2012). Virtual screening: An emergent, key methodology for drug development in an emergent continent: A bridge towards patentability. In E. Castro & A. Haghi (Eds.), *Advanced methods and applications in chemoinformatics: Research progress and new applications* (pp. 229–245). Hershey, PA: IGI Global. doi:10.4018/978-1-60960-860-6.ch011

Tauber, A. (2012). Requirements and properties of qualified electronic delivery systems in egovernment: An Austrian experience. In S. Sharma (Ed.), *E-adoption and technologies for empowering developing countries: Global advances* (pp. 115–128). Hershey, PA: IGI Global. doi:10.4018/978-1-4666-0041-6.ch009

Telesko, R., & Nikles, S. (2012). Semantic-enabled compliance management. In S. Smolnik, F. Teuteberg, & O. Thomas (Eds.), *Semantic technologies for business and information systems engineering: Concepts and applications* (pp. 292–310). Hershey, PA: IGI Global. doi:10.4018/978-1-60960-126-3.ch015

Tella, A., & Afolabi, A. K. (2013). Internet policy issues and digital libraries' management of intellectual property. In S. Thanuskodi (Ed.), *Challenges of academic library management in developing countries* (pp. 272–284). Hershey, PA: IGI Global. doi:10.4018/978-1-4666-4070-2.ch019

Tiwari, S. C., Gupta, M., Khan, M. A., & Ansari, A. Q. (2013). Intellectual property rights in semi-conductor industries: An Indian perspective. In S. Saeed, M. Khan, & R. Ahmad (Eds.), *Business strategies and approaches for effective engineering management* (pp. 97–110). Hershey, PA: IGI Global. doi:10.4018/978-1-4666-3658-3.ch006

*Related References*

Truyen, F., & Buekens, F. (2013). Professional ICT knowledge, epistemic standards, and social epistemology. In T. Takševa (Ed.), *Social software and the evolution of user expertise: Future trends in knowledge creation and dissemination* (pp. 274–294). Hershey, PA: IGI Global. doi:10.4018/978-1-4666-2178-7.ch016

Tummons, J. (2011). Deconstructing professionalism: An actor-network critique of professional standards for teachers in the UK lifelong learning sector. *International Journal of Actor-Network Theory and Technological Innovation, 3*(4), 22–31. doi:10.4018/jantti.2011100103

Tummons, J. (2013). Deconstructing professionalism: An actor-network critique of professional standards for teachers in the UK lifelong learning sector. In A. Tatnall (Ed.), *Social and professional applications of actor-network theory for technology development* (pp. 78–87). Hershey, PA: IGI Global. doi:10.4018/978-1-4666-2166-4.ch007

Tuohey, W. G. (2014). Lessons from practices and standards in safety-critical and regulated sectors. In I. Ghani, W. Kadir, & M. Ahmad (Eds.), *Handbook of research on emerging advancements and technologies in software engineering* (pp. 369–391). Hershey, PA: IGI Global. doi:10.4018/978-1-4666-6026-7.ch016

Tzoulia, E. (2013). Legal issues to be considered before setting in force consumer-centric marketing strategies within the European Union. In H. Kaufmann & M. Panni (Eds.), *Customer-centric marketing strategies: Tools for building organizational performance* (pp. 36–56). Hershey, PA: IGI Global. doi:10.4018/978-1-4666-2524-2.ch003

Unland, R. (2012). Interoperability support for e-business applications through standards, services, and multi-agent systems. In E. Kajan, F. Dorloff, & I. Bedini (Eds.), *Handbook of research on e-business standards and protocols: Documents, data and advanced web technologies* (pp. 129–153). Hershey, PA: IGI Global. doi:10.4018/978-1-4666-0146-8.ch007

Uslar, M., Grüning, F., & Rohjans, S. (2013). A use case for ontology evolution and interoperability: The IEC utility standards reference framework 62357. In M. Khosrow-Pour (Ed.), *Cases on performance measurement and productivity improvement: Technology integration and maturity* (pp. 387–415). Hershey, PA: IGI Global. doi:10.4018/978-1-4666-2618-8.ch018 van de Kaa, G. (2013). Responsible innovation and standardization: A new research approach? *International Journal of IT Standards and Standardization Research*, *11*(2), 61–65. doi:10.4018/jitsr.2013070105

van de Kaa, G., Blind, K., & de Vries, H. J. (2013). The challenge of establishing a recognized interdisciplinary journal: A citation analysis of the international journal of IT standards and standardization research. *International Journal of IT Standards and Standardization Research*, *11*(2), 1–16. doi:10.4018/jitsr.2013070101

Venkataraman, H., Ciubotaru, B., & Muntean, G. (2012). System design perspective: WiMAX standards and IEEE 802.16j based multihop WiMAX. In G. Cornetta, D. Santos, & J. Vazquez (Eds.), *Wireless radio-frequency standards and system design: Advanced techniques* (pp. 287–309). Hershey, PA: IGI Global. doi:10.4018/978-1-4666-0083-6.ch012

Vishwakarma, P., & Mukherjee, B. (2014). Knowing protection of intellectual contents in digital era. In N. Patra, B. Kumar, & A. Pani (Eds.), *Progressive trends in electronic resource management in libraries* (pp. 147–165). Hershey, PA: IGI Global. doi:10.4018/978-1-4666-4761-9.ch008

Wasilko, P. J. (2011). Law, architecture, gameplay, and marketing. In M. Cruz-Cunha, V. Varvalho, & P. Tavares (Eds.), *Business, technological, and social dimensions of computer games: Multidisciplinary developments* (pp. 476–493). Hershey, PA: IGI Global. doi:10.4018/978-1-60960-567-4.ch029

Wasilko, P. J. (2012). Law, architecture, gameplay, and marketing. In *Computer engineering: concepts, methodologies, tools and applications* (pp. 1660–1677). Hershey, PA: IGI Global. doi:10.4018/978-1-61350-456-7.ch703

Wasilko, P. J. (2014). Beyond compliance: Understanding the legal aspects of information system administration. In I. Portela & F. Almeida (Eds.), *Organizational, legal, and technological dimensions of information system administration* (pp. 57–75). Hershey, PA: IGI Global. doi:10.4018/978-1-4666-4526-4.ch004

White, G. L., Mediavilla, F. A., & Shah, J. R. (2011). Information privacy: Implementation and perception of laws and corporate policies by CEOs and managers. *International Journal of Information Security and Privacy*, *5*(1), 50–66. doi:10.4018/jisp.2011010104

*Related References*

White, G. L., Mediavilla, F. A., & Shah, J. R. (2013). Information privacy: Implementation and perception of laws and corporate policies by CEOs and managers. In H. Nemati (Ed.), *Privacy solutions and security frameworks in information protection* (pp. 52–69). Hershey, PA: IGI Global. doi:10.4018/978-1-4666-2050-6.ch004

Whyte, K. P., List, M., Stone, J. V., Grooms, D., Gasteyer, S., Thompson, P. B., & Bouri, H. et al. (2014). Uberveillance, standards, and anticipation: A case study on nanobiosensors in U.S. cattle. In M. Michael & K. Michael (Eds.), *Uberveillance and the social implications of microchip implants: Emerging technologies* (pp. 260–279). Hershey, PA: IGI Global. doi:10.4018/978-1-4666-4582-0.ch012

Wilkes, W., Reusch, P. J., & Moreno, L. E. (2012). Flexible classification standards for product data exchange. In E. Kajan, F. Dorloff, & I. Bedini (Eds.), *Handbook of research on e-business standards and protocols: Documents, data and advanced web technologies* (pp. 448–466). Hershey, PA: IGI Global. doi:10.4018/978-1-4666-0146-8.ch021

Wittkower, D. E. (2011). Against strong copyright in e-business. In *Global business: Concepts, methodologies, tools and applications* (pp. 2157–2176). Hershey, PA: IGI Global. doi:10.4018/978-1-60960-587-2.ch720

Wright, D. (2012). Evolution of standards for smart grid communications. *International Journal of Interdisciplinary Telecommunications and Networking, 4*(1), 47–55. doi:10.4018/jitn.2012010103

Wurster, S. (2013). Development of a specification for data interchange between information systems in public hazard prevention: Dimensions of success and related activities identified by case study research. *International Journal of IT Standards and Standardization Research, 11*(1), 46–66. doi:10.4018/jitsr.2013010103

Wyburn, M. (2011). Copyright and ethical issues in emerging models for the digital media reporting of sports news in Australia. In M. Quigley (Ed.), *ICT ethics and security in the 21st century: New developments and applications* (pp. 66–85). Hershey, PA: IGI Global. doi:10.4018/978-1-60960-573-5.ch004

Wyburn, M. (2013). Copyright and ethical issues in emerging models for the digital media reporting of sports news in Australia. In *Digital rights management: Concepts, methodologies, tools, and applications* (pp. 290–309). Hershey, PA: IGI Global. doi:10.4018/978-1-4666-2136-7.ch014

Xiaohui, T., Yaohui, Z., & Yi, Z. (2012). The management system of enterprises intellectual property rights: A case study from China. *International Journal of Asian Business and Information Management, 3*(1), 50–64. doi:10.4018/jabim.2012010105

Xiaohui, T., Yaohui, Z., & Yi, Z. (2013). The management system of enterprises' intellectual property rights: A case study from China. In *Digital rights management: Concepts, methodologies, tools, and applications* (pp. 1092–1106). Hershey, PA: IGI Global. doi:10.4018/978-1-4666-2136-7.ch053

Xuan, X., & Xiaowei, Z. (2012). The dilemma and resolution: The patentability of traditional Chinese medicine. *International Journal of Asian Business and Information Management, 3*(3), 1–8. doi:10.4018/jabim.2012070101

Yang, C., & Lu, Z. (2011). A blind image watermarking scheme utilizing BTC bitplanes. *International Journal of Digital Crime and Forensics, 3*(4), 42–53. doi:10.4018/jdcf.2011100104

Yastrebenetsky, M., & Gromov, G. (2014). International standard bases and safety classification. In M. Yastrebenetsky & V. Kharchenko (Eds.), *Nuclear power plant instrumentation and control systems for safety and security* (pp. 31–60). Hershey, PA: IGI Global. doi:10.4018/978-1-4666-5133-3.ch002

Zouag, N., & Kadiri, M. (2014). Intellectual property rights, innovation, and knowledge economy in Arab countries. In A. Driouchi (Ed.), *Knowledge-based economic policy development in the Arab world* (pp. 245–272). Hershey, PA: IGI Global. doi:10.4018/978-1-4666-5210-1.ch010

# Compilation of References

Agarwal, D. (2007). Detecting anomalies in cross-classified streams: A bayesian approach. *Knowledge and Information Systems, 11*(1), 29–44. doi:10.1007/s10115-006-0036-4

Aggarwal, C. C. (2013). *Outlier Analysis.* New York: Springer. doi:10.1007/978-1-4614-6396-2

Ahmed, T., Oreshkin, B., & Coates, M. J. (2007). Machine learning approaches to network anomaly detection. In *SYSML'07 Proceedings of the 2ⁿᵈ USENIX workshop on Tackling computer systems problems with machine learning techniques.* Berkeley, CA: USENIX Association.

Ahmed, T., Oreshkin, B., & Coates, M. J. (2007). April). Machine learning approaches to network anomaly detection. In *Proceedings of the 2nd USENIX workshop on Tackling computer systems problems with machine learning techniques* (pp. 1-6). USENIX Association.

Alan, D. S., & Basu, K. (1994). Customer loyalty: Toward an integrated conceptual framework. *Journal of the Academy of Marketing Science, 22*(2), 99–113. doi:10.1177/0092070394222001

Aliu, O. G., Imran, A., Imran, M. A., & Evans, B. (2013). A survey of self organisation in future cellular networks. *IEEE Communications Surveys and Tutorials, 15*(1), 336–361. doi:10.1109/SURV.2012.021312.00116

Amzallag, D., Bar-Yehuda, R., Raz, D., & Scalosub, G. (2013). Cell selection in 4G cellular networks. *IEEE Transactions on Mobile Computing, 12*(7), 1443–1455. doi:10.1109/TMC.2012.83

Aral, S., Muchnik, L., & Sundararajan, A. (2009). Distinguishing influence based contagion from homophily-driven diffusion in dynamic networks. *Proceedings of the National Academy of Sciences of the United States of America*, *106*(51), 21544–21549. doi:10.1073/pnas.0908800106 PMID:20007780

Au, W. H., Chan, K. C., & Yao, X. (2003). A novel evolutionary data mining algorithm with applications to churn prediction. *IEEE Transactions on Evolutionary Computation*, *7*(6), 532–545. doi:10.1109/TEVC.2003.819264

Bamnett, V., & Lewis, T. (1994). *Outliers in statistical data*. Chichester, UK: John Wiley & Sons.

Barfiord, P., Kline, J., Plonka, D., & Ron, A. (2002). A signal analysis of network traffic anomalies. In *Proceedings of ACM SIGCOMM Internet Measurement Workshop*. New York: ACM. doi:10.1145/637201.637210

Barlow, H. B. (1989). Unsupervised learning. *Neural Computation*, *1*(3), 295–311. doi:10.1162/neco.1989.1.3.295

Bass, F. M. (1969). A new product growth model for consumer durables. *Management Science*, *15*(1), 215–227. doi:10.1287/mnsc.15.5.215

Berry, M. J., & Linoff, G. S. (2004). *Data mining techniques: for marketing, sales, and customer relationship management*. John Wiley & Sons.

Bezdek, J. C. (2013). *Pattern recognition with fuzzy objective function algorithms*. Berlin, Germany: Springer Science & Business Media.

Bin, L., Peiji, S., & Juan, L. (2007, June). Customer churn prediction based on the decision tree in personal handyphone system service. In *Service Systems and Service Management, 2007 International Conference on* (pp. 1-5). IEEE. doi:10.1109/ICSSSM.2007.4280145

Borshchev, A., & Filippov, A. (2004, July). From system dynamics and discrete event to practical agent based modeling: reasons, techniques, tools.*Proceedings of the 22nd international conference of the system dynamics society*, 22.

Bozdogan, H. (1987). Model selection and Akaikes information criterion (AIC): The general theory and its analytical extensions. *Psychometrika*, *52*(3), 345–370. doi:10.1007/BF02294361

Breiman, L., Friedman, J., Stone, C. J., & Olshen, R. A. (1984). *Classification and regression trees*. CRC Press.

Brin, S., & Page, L. (1998). The anatomy of a large-scale hypertextual Web search engine. *Computer Networks and ISDN Systems, 30*(1-7), 107–117. doi:10.1016/S0169-7552(98)00110-X

Brutlag, J. D. (2000). Aberrant behavior detection in time series for network service monitoring. In *Proceeding of the 14th Systems Administration Conference* (pp. 139-146). USENIX Association.

Burges, C. J. C. (2010). From ranknet to lambdarank to lambdamart: An overview. *Learning, 11*, 23–581.

Burnham, K. P., & Anderson, D. R. (2004). Multimodel inference understanding AIC and BIC in model selection. *Sociological Methods & Research, 33*(2), 261–304. doi:10.1177/0049124104268644

Burt, R. S. (1987). Social contagion and innovation: Cohesion versus structural equivalence. *American Journal of Sociology, 92*(6), 1287–1335. doi:10.1086/228667

Cantono, S., & Silverberg, G. (2009). A percolation model of eco-innovation diffusion: The relationship between diffusion, learning economies and subsidies. *Technological Forecasting and Social Change, 76*(4), 487–496. doi:10.1016/j.techfore.2008.04.010

Chandola, V., Banerjee, A., & Kumar, V. (2009). Anomaly detection: A survey. *ACM Computing Surveys, 41*(3), 15. doi:10.1145/1541880.1541882

Chawla, N. V., Japkowicz, N., & Kotcz, A. (2004). Editorial: Special issue on learning from imbalanced data sets. *ACM Sigkdd Explorations Newsletter, 6*(1), 1–6. doi:10.1145/1007730.1007733

Chen, M. S., Han, J., & Yu, P. S. (1996). Data mining: An overview from a database perspective. *Knowledge and data Engineering. IEEE Transactions on, 8*(6), 866–883.

Chevalier, J. A., & Mayzlin, D. (2006). The effect of word of mouth on sales: Online book reviews. *JMR, Journal of Marketing Research, 43*(3), 345–354. doi:10.1509/jmkr.43.3.345

Chiang, M. M. T., & Mirkin, B. (2010). Intelligent choice of the number of clusters in k-means clustering: an experimental study with different cluster spreads. *Journal of Classification, 27*(1), 3-40.

Clauset, A., Newman, M. E. J., & Moore, C. (2005). Finding community structure in very large networks. *Physical Review E: Statistical, Nonlinear, and Soft Matter Physics*, *70*(6), 066111. doi:10.1103/PhysRevE.70.066111 PMID:15697438

Cohen, R., & Havlin, S. (2010). *Complex Networks: Structure, Robustness and Function*. Cambridge University Press. doi:10.1017/CBO9780511780356

Coleman, J. S., Katz, E., & Menzel, H. (1966). *Medical innovation: A diffusion study*. Bobbs-Merrill Co.

Cortes, C., & Mohri, M. (2004). AUC optimization vs. error rate minimization. *Advances in Neural Information Processing Systems*, *16*, 313–320.

Dasgupta, K., Singh, R., Viswanathan, B., Chakraborty, D., Mukherjea, S., Nanavati, A. A., & Joshi, A. (2008, March). Social ties and their relevance to churn in mobile telecom networks. In *Proceedings of the 11th international conference on Extending database technology: Advances in database technology* (pp. 668-677). ACM. doi:10.1145/1353343.1353424

Dempster, A. P., Laird, N. M., & Rubin, D. B. (1977). Maximum likelihood from incomplete data via the EM algorithm. *Journal of the Royal Statistical Society. Series B. Methodological*, *39*, 1–38.

Dempster, A., Laird, N., & Rubin, D. (1977). Maximum Likelihood from Incomplete Data via the EM Algorithm. *Journal of the Royal Statistical Society. Series A (General)*, *39*(1), 1–38.

Desforges, M. J., Jacob, P. J., & Cooper, J. E. (1998). Applications of probability density estimation to the detection of abnormal conditions in engineering. *Proceedings - Institution of Mechanical Engineers*, *212*(8), 687–703.

Desforges, M. J., Jacob, P. J., & Cooper, J. E. (1998). Applications of probability density estimation to the detection of abnormal conditions in engineering. *Proceedings of the Institution of Mechanical Engineers. Part C, Journal of Mechanical Engineering Science*, *212*(8), 687–703. doi:10.1243/0954406981521448

Dharmaraja, S., Jindal, V., & Varshney, U. (2008). Reliability and survivability analysis for UMTS networks: An analytical approach. *IEEE eTransactions on Network and Service Management*, *5*(3), 132–142. doi:10.1109/TNSM.2009.031101

Dietterich, T. G. (2000). Ensemble methods in machine learning. In J. Kittler & F. Roli (Eds.), *Multiple classifier systems* (pp. 1–15). Berlin: Springer. doi:10.1007/3-540-45014-9_1

Domingos, P. (1999). The role of Occams razor in knowledge discovery. *Data Mining and Knowledge Discovery*, *3*(4), 409–425. doi:10.1023/A:1009868929893

Eiben, A. E., Koudijs, A. E., & Slisser, F. (1998). Genetic modelling of customer retention. In *Genetic Programming* (pp. 178–186). Springer Berlin Heidelberg. doi:10.1007/BFb0055937

Engels, A., Reyer, M., Xu, X., Mathar, R., Zhang, J., & Zhuang, H. (2013). Autonomous self-optimization of coverage and capacity in LTE cellular networks. *IEEE Transactions on Vehicular Technology*, *62*(5), 1989–2004. doi:10.1109/TVT.2013.2256441

Entner, R. (2011, June 23). *International comparisons: the handset replacement cycle*. Retrieved from http://mobilefuture.org/wp-content/uploads/2013/02/mobile-future.publications.handset-replacement-cycle.pdf

Eskin, E., Arnold, A., Prerau, M., Portnoy, L., & Stolfo, S. (2002). A geometric framework for unsupervised anomaly detection. In Applications of data mining in computer security (pp. 77-101). Springer US. doi:10.1007/978-1-4615-0953-0_4

European Telecommunications Standards Institute. (2014). *Requirements for Further Advancements for Evolved Universal Terrestrial Radio Access*. Retrieved from http://www.etsi.org/deliver/etsi_tr/136900_136999/136913/10.00.00_60/tr_136913v100000p.pdf

Fibich, G., & Gibori, R. I. (2010). Aggregate diffusion dynamics in agent-based models with a spatial structure. *Operations Research*, *58*(5), 1450–1468. doi:10.1287/opre.1100.0818

Fisher, J. C., & Pry, R. H. (1972). A simple substitution model of technological change. *Technological Forecasting and Social Change*, *3*, 75–88. doi:10.1016/S0040-1625(71)80005-7

Fourt, L. A., & Woodlock, J. W. (1960). Early prediction of market success for new grocery products. *Journal of Marketing*, *25*(2), 31–38. doi:10.2307/1248608

Fowler, F. J. J. (2013). *Survey research methods*. Thousand Oaks, CA: Sage publications.

Fred, A. (2001). Finding consistent clusters in data partitions. In *MCS'01 Proceedings of the Second International Workshop on Multiple classifier systems* (pp. 309-318). London, UK: Springer. doi:10.1007/3-540-48219-9_31

Friedman, J. H. (1991). Multivariate adaptive regression splines. *Annals of Statistics*, *19*(1), 1–67. doi:10.1214/aos/1176347963

Gobjuka, H. (2009). *4G wireless networks: Opportunities and challenges*. arXiv preprint arXiv: 0907.2929

Godinho de Matos, M., Ferreira, P., & Krackhardt, D. (2014). (Forthcoming). Peer Influence in the Diffusion of the iPhone 3G over a Large Social Network. *Management Information Systems Quarterly*.

Goldenberg, J., Libai, B., & Muller, E. (2001). Using complex systems analysis to advance marketing theory development: Modeling heterogeneity effects on new product growth through stochastic cellular automata. *Academy of Marketing Science Review*, *9*(3), 1–18.

Goodman, J., & Newman, S. (2003). Understand customer behavior and complaints. *Quality Progress*, *36*(1), 51–55.

Goodman, L. A. (1961). Snowball sampling. *Annals of Mathematical Statistics*, *32*(1), 148–170. doi:10.1214/aoms/1177705148

Grauwin, S., Sobolevsky, S., Moritz, S., Gódor, I., & Ratti, C. (2015). Towards a comparative science of cities: Using mobile traffic records in new york, london, and hong kong. In *Computational approaches for urban environments* (pp. 363–387). Springer International Publishing. doi:10.1007/978-3-319-11469-9_15

Guerrero, V. M., & Johnson, R. A. (1982). Use of the Box-Cox transformation with binary response models. *Biometrika*, *69*(2), 309–314. doi:10.1093/biomet/69.2.309

Guida, M., Longo, M., & Postiglione, F. (2010, December). Performance evaluation of IMS-based core networks in presence of failures. In *Global Telecommunications Conference (GLOBECOM 2010)*, (pp. 1-5). IEEE. doi:10.1109/GLOCOM.2010.5683540

Hadden, J., Tiwari, A., Roy, R., & Ruta, D. (2006). Churn prediction using complaints data. *Proceedings of World Academy of Science, Engineering and Technology*.

Hajji, H. (2005). Statistical analysis of network traffic for adaptive faults detection. *IEEE Transactions on Neural Networks*, *16*(5), 1053–1063. doi:10.1109/TNN.2005.853414 PMID:16252816

Halbauer, H., Saur, S., Koppenborg, J., & Hoek, C. (2012, April). Interference avoidance with dynamic vertical beamsteering in real deployments. In Wireless Communications and Networking Conference Workshops (WCNCW), 2012 IEEE (pp. 294-299). IEEE. doi:10.1109/WCNCW.2012.6215509

Hämäläinen, S., Sanneck, H., & Sartori, C. (2012). *LTE self-organising networks (SON): network management automation for operational efficiency*. John Wiley & Sons.

Hand, D. J. (2009). Measuring classifier performance: A coherent alternative to the area under the ROC curve. *Machine Learning, 77*(1), 103–123. doi:10.1007/s10994-009-5119-5

Hand, D. J., & Anagnostopoulos, C. (2013). When is the area under the receiver operating characteristic curve an appropriate measure of classifier performance? *Pattern Recognition Letters, 34*(5), 492–495. doi:10.1016/j.patrec.2012.12.004

Hastie, T. J., & Tibshirani, R. J. (1990). *Generalized additive models* (Vol. 43). CRC Press.

Hebb, D. (1949). *The Organization of Behavior*. New York: Wiley.

Hinton, G. E., & Sejnowski, T. J. (1986). Learning and relearning in Boltzmann machines. In Parallel Distributed Processing: Explorations in the Microstructure of Cognition:*Foundations* (vol. 1, pp. 282–317). Cambridge, MA: MIT Press.

Hofmann, T. (1999a). Probabilistic latent semantic indexing. In *Proceedings of the 22nd annual international ACM SIGIR conference on Research and development in information retrieval (SIGIR '99)* (pp. 50-54). New York, NY: ACM. doi:10.1145/312624.312649

Hofmann, T. (1999b). Probabilistic latent semantic analysis. In *Proceedings of the Fifteenth conference on Uncertainty in artificial intelligence* (pp. 289-296). Burlington, MA: Morgan Kaufmann Publishers Inc.

Hohnisch, M., Pittnauer, S., & Stauffer, D. (2008). A percolation-based model explaining delayed takeoff in new-product diffusion. *Industrial and Corporate Change, 17*(5), 1001–1017. doi:10.1093/icc/dtn031

Hong Kong Government. (2015). *Hong Kong: The Facts*. Retrieved from http://www.gov.hk/en/about/abouthk/factsheets/docs/telecommunications.pdf

Hosmer, J. D. W., & Lemeshow, S. (2004). *Applied logistic regression*. New York: John Wiley & Sons.

Hu, M., Hsieh, C., & Jia, J. (2014). *The effectiveness of peer influence and network structure: an application using mobile data*. Working Paper.

Hu, M., Yang, S., & Xu, Y. (2015). *Social Learning and Network Effects in Contagious Switching Behavior.* Working Paper.

Hu, S., Ouyang, Y., Yao, Y., Fallah, M. H., & Lu, W. (2014). A study of LTE network performance based on data analytics and statistical modeling. In *2014 23rd Wireless and Optical Communication Conference (WOCC)* (pp. 1-6). IEEE.

Huang, J., & Ling, C. X. (2005). Using AUC and accuracy in evaluating learning algorithm. *IEEE Transactions on Knowledge and Data Engineering, 17*(3), 299–310. doi:10.1109/TKDE.2005.50

Huang, J., & Ling, C. X. (2005). Using AUC and accuracy in evaluating learning algorithms. *Knowledge and Data Engineering. IEEE Transactions, 17*(3), 299–310.

Hung, S. Y., Yen, D. C., & Wang, H. Y. (2006). Applying data mining to telecom churn management. *Expert Systems with Applications, 31*(3), 515–524. doi:10.1016/j.eswa.2005.09.080

Hurvich, C. M., & Tsai, C. (1989). Regression and time series model selection in small samples. *Biometrika, 76*(2), 297–307. doi:10.1093/biomet/76.2.297

Imran, A., Imran, M. A., & Tafazolli, R. (2011, December). Distributed spectral efficiency optimization at hotspots through self organisation of BS tilts. In 2011 IEEE GLOBECOM Workshops (GC Wkshps) (pp. 570-574). IEEE. doi:10.1109/GLOCOMW.2011.6162515

Imran, A., Imran, M. A., Abu-Dayya, A., & Tafazolli, R. (2014). Self organization of tilts in relay enhanced networks: A distributed solution. *IEEE Transactions on Wireless Communications, 13*(2), 764–779. doi:10.1109/TWC.2014.011614.130299

Imran, A., Zoha, A., & Abu-Dayya, A. (2014). Challenges in 5G: How to empower SON with big data for enabling 5G. *IEEE Network, 28*(6), 27–33. doi:10.1109/MNET.2014.6963801

Imran, M. A., Imran, A., & Tafazolli, R. (2011). Relay station access link spectral efficiency optimization through SO of macro BS tilts. *IEEE Communications Letters, 15*(12), 1326–1328. doi:10.1109/LCOMM.2011.103111.1579

International Data Corporation. (2015). *Smartphone Vendor Market Share, Q12015.* Retrieved from http://www.idc.com/prodserv/smartphone-market-share.jsp

International Telecommunication Union. (2015). *World 2015.* Retrieved from http://www.itu.int/en/ITU-D/Statistics/Documents/facts/ICTFactsFigures2015.pdf

Iyengar, R., Van den Bulte, C., & Lee, J. Y. (2015). Social contagion in new product trial and repeat. *Marketing Science*, *34*(3), 408–429. doi:10.1287/mksc.2014.0888

Iyengar, R., Van den Bulte, C., & Valente, T. W. (2011). Opinion leadership and social contagion in new product diffusion. *Marketing Science*, *30*(2), 195–212. doi:10.1287/mksc.1100.0566

Jain, A. K. (2010). Data clustering: 50 years beyond K-means. *Pattern Recognition Letters*, *31*(8), 651–666. doi:10.1016/j.patrec.2009.09.011

Jain, A. K., Murty, M. N., & Flynn, P. J. (1999). Data clustering: A review. *ACM Computing Surveys*, *31*(3), 264–323. doi:10.1145/331499.331504

Kass, G. V. (1980). An exploratory technique for investigating large quantities of categorical data. *Applied Statistics*, *29*(2), 119–127. doi:10.2307/2986296

Katz, E., & Lazarsfeld, P. F. (1955). *Personal Influence, The part played by people in the flow of mass communications*. Transcation Publishers.

Katz, M. L., & Shapiro, C. (1985). Network externalities, competition, and compatibility. *The American Economic Review*, 424–440.

Kaufman, L., & Rousseeuw, P. (1987). *Clustering by means of medoids*. Delft, The Netherlands: North-Holland.

Kaufman, L., & Rousseeuw, P. J. (2009). *Finding groups in data: an introduction to cluster analysis* (Vol. 344). New York: John Wiley & Sons.

Khan, A., Sun, L., & Ifeachor, E. (2012). QoE prediction model and its application in video quality adaptation over UMTS networks. *IEEE Transactions on Multimedia*, *14*(2), 431–442. doi:10.1109/TMM.2011.2176324

Kiesling, E., Günther, M., Stummer, C., & Wakolbinger, L. M. (2012). Agent-based simulation of innovation diffusion: A review. *Central European Journal of Operations Research*, *20*(2), 183–230. doi:10.1007/s10100-011-0210-y

Kisioglu, P., & Topcu, Y. I. (2011). Applying Bayesian Belief Network approach to customer churn analysis: A case study on the telecom industry of Turkey. *Expert Systems with Applications*, *38*(6), 7151–7157. doi:10.1016/j.eswa.2010.12.045

Kocsis, G., & Kun, F. (2008). The effect of network topologies on the spreading of technological developments. *Journal of Statistical Mechanics*, *2008*(10), P10014. doi:10.1088/1742-5468/2008/10/P10014

Kohonen, T. (1987). Adaptive, associative, and self-organizing functions in neural computing. *Applied Optics*, *26*(23), 4910–4918. doi:10.1364/AO.26.004910 PMID:20523469

Kuhn, M. (2008). Building predictive models in R using the caret package. *Journal of Statistical Software*, *28*(5), 1–26. doi:10.18637/jss.v028.i05 PMID:27774042

Kulkarni, V. G. (2011). *Introduction to modelling and analysis of stochastic systems*. Springer.

Lam, H. W., & Wu, C. (2009). Finding influential ebay buyers for viral marketing a conceptual model of BuyerRank. In *2009 International Conference on Advanced Information Networking and Applications* (pp. 778-785). IEEE. doi:10.1109/AINA.2009.36

Langley, P., Iba, W., & Thompson, K. (1992, July). *An analysis of Bayesian classifiers* (Vol. 90). AAAI.

Laskov, P., Düssel, P., Schäfer, C., & Rieck, K. (2005). *Learning intrusion detection: supervised or unsupervised? In Image Analysis and Processing–ICIAP 2005* (pp. 50–57). Berlin: Springer. doi:10.1007/11553595_6

Lateef, H. Y., Imran, A., & Abu-Dayya, A. (2013, September). A framework for classification of Self-Organising network conflicts and coordination algorithms. In *2013 IEEE 24th Annual International Symposium on Personal, Indoor, and Mobile Radio Communications (PIMRC)* (pp. 2898-2903). IEEE. doi:10.1109/PIMRC.2013.6666642

Lazarov, V., & Capota, M. (2007). *Churn prediction*. Bus. Anal. Course. TUM Comput. Sci.

Lee, K., Lee, H., Jang, Y. U., & Cho, D. H. (2013). CoBRA: Cooperative beamforming-based resource allocation for self-healing in SON-based indoor mobile communication system. *IEEE Transactions on Wireless Communications*, *12*(11), 5520–5528. doi:10.1109/TWC.2013.092013.121429

Leenders, A. J. (2002). Modeling social influence through network autocorrelation: Constructing the weight matrix. *Social Networks*, *24*(1), 21–47. doi:10.1016/S0378-8733(01)00049-1

Lee, Y. J., Hosanagar, K., & Tan, Y. (2015). Do I follow my friends or the crowd? Information cascades in online movie ratings. *Management Science*, *61*(9), 2241–2258. doi:10.1287/mnsc.2014.2082

Lemmens, A., & Croux, C. (2006). Bagging and boosting classification trees to predict churn. *JMR, Journal of Marketing Research, 43*(2), 276–286. doi:10.1509/jmkr.43.2.276

Leung, K., & Leckie, C. (2005). Unsupervised anomaly detection in network intrusion detection using clusters. In *Proceedings of the Twenty-eighth Australasian conference on Computer Science* (Vol. 38, pp. 333-342). Australian Computer Society, Inc.

Liaw, A., & Wiener, M. (2002). Classification and regression by RandomForest. *R News, 2*(3), 18-22.

Liaw, A., & Wiener, M. (2002). Classification and Regression by randomForest. *R News, 2*(3), 18-22.

Liebig, J., & Rao, A. (2014). Identifying Influential Nodes in Bipartite Networks Using the Clustering Coefficient. In *2014 Tenth International Conference on Signal-Image Technology and Internet-Based Systems (SITIS)* (pp. 323-330). IEEE. doi:10.1109/SITIS.2014.15

Lobo, J. M., Jiménez-Valverde, A., & Real, R. (2008). AUC: A misleading measure of the performance of predictive distribution models. *Global Ecology and Biogeography, 17*(2), 145–151. doi:10.1111/j.1466-8238.2007.00358.x

Lu, Z., Zhong, E., Zhao, L., Xiang, E., Pan, W., & Yang, Q. (2013). *Selective Transfer Learning for Cross Domain Recommendation*. Philadelphia, PA: SDM. doi:10.14711/thesis-b1240240

MacCartney, G. R., Zhang, J., Nie, S., & Rappaport, T. S. (2013, December). Path loss models for 5G millimeter wave propagation channels in urban microcells. In *2013 IEEE Global Communications Conference (GLOBECOM)*(pp. 3948-3953). IEEE. doi:10.1109/GLOCOM.2013.6831690

MacKay, D. M. (1956). The epistemological problem for automata. In C. E. Shannon & J. McCarthy (Eds.), *Automata Studies* (pp. 235–251). Princeton, NJ: Princeton University Press.

MacQueen, J. (1967). Some methods for classification and analysis of multivariate observations. In *Proceedings of the fifth Berkeley symposium on mathematical statistics and probability* (Vol. 1, pp. 281-297). Oakland, CA: University of California Press.

Mahajan, V., & Muller, E. (1979). Innovation diffusion and new product growth models in marketing. *Journal of Marketing, 43*(4), 55–68. doi:10.2307/1250271

Mahajan, V., Muller, E., & Bass, F. M. (1990). New product diffusion models in marketing: A review and directions for research. *Journal of Marketing, 54*(1), 1–26. doi:10.2307/1252170

Manchanda, P., Xie, Y., & Youn, N. (2008). The role of targeted communication and contagion in product adoption. *Marketing Science, 27*(6), 961–976. doi:10.1287/mksc.1070.0354

Mansfield, E. (1961). Technical change and the rate of imitation. *Econometrica, 29*(4), 741–766. doi:10.2307/1911817

Marr, D. (1970). A theory for cerebral neocortex. *Proceedings of the Royal Society of London. Series B, Biological Sciences, 176*(1043), 161–234. doi:10.1098/rspb.1970.0040 PMID:4394740

Masand, B., Datta, P., Mani, D. R., & Li, B. (1999). CHAMP: A prototype for automated cellular churn prediction. *Data Mining and Knowledge Discovery, 3*(2), 219–225. doi:10.1023/A:1009873905876

McLachlan, G. J., & Basford, K. E. (1988). *Mixture models. Inference and applications to clustering.* New York: Dekker.

McLachlan, G. J., & Basford, K. E. (1988). *Mixture models: Inference and applications to clustering.* New York: Dekker.

Menzel, H. (1960). Innovation, integration, and marginality: A survey of physicians. *American Sociological Review, 25*(5), 704–713. doi:10.2307/2090143

Miller, A. (2002). *Subset selection in regression.* CRC Press. doi:10.1201/9781420035933

Moretti, E. (2011). Social learning and peer effects in consumption: Evidence from movie sales. *The Review of Economic Studies, 78*(1), 356–393. doi:10.1093/restud/rdq014

Muchnik, L., Aral, S., & Taylor, S. J. (2013). Social influence bias: A randomized experiment. *Science, 341*(6146), 647–651. doi:10.1126/science.1240466 PMID:23929980

*Compilation of References*

Muhammad, N. U. I., Abou-Jaoude, R., Hartmann, C., & Mitschele-Thiel, A. (2010, May). Self-Optimization of Antenna Tilt and Pilot Power for Dedicated Channels. *WiOpt'10: Modeling and Optimization in Mobile, Ad Hoc, and Wireless Networks,* 278-285.

Nambiar, R., & Poess, M. (2011). Transaction performance vs. Moore's law: a trend analysis. In Performance Evaluation, Measurement and Characterization of Complex Systems (pp. 110-120). Springer Berlin Heidelberg.

Narayan, V., Rao, V. R., & Saunders, C. (2011). How peer influence affects attribute preferences: A Bayesian updating mechanism. *Marketing Science, 30*(2), 368–384. doi:10.1287/mksc.1100.0618

Nath, S. V., & Behara, R. S. (2003, November). Customer churn analysis in the wireless industry: A data mining approach. *Proceedings-annual meeting of the decision sciences institute,* 505-510.

Navaie, K., & Sharafat, A. R. (2003). A framework for UMTS air interface analysis. *Canadian Journal of Electrical and Computer Engineering, 28*(3/4), 113–129. doi:10.1109/CJECE.2003.1425098

Neslin, S. A., Gupta, S., Kamakura, W., Lu, J., & Mason, C. H. (2006). Defection detection: Measuring and understanding the predictive accuracy of customer churn models. *JMR, Journal of Marketing Research, 43*(2), 204–211. doi:10.1509/jmkr.43.2.204

Newman, M. E. J. (2003). The structure and function of complex networks. *SIAM Review, 45*(2), 167–256. doi:10.1137/S003614450342480

Newman, M. E. J. (2004). Detecting community structure in networks. *The European Physical Journal B-Condensed Matter and Complex Systems, 38*(2), 321–330. doi:10.1140/epjb/e2004-00124-y PMID:15244693

Ngai, E. W., Xiu, L., & Chau, D. C. (2009). Application of data mining techniques in customer relationship management: A literature review and classification. *Expert Systems with Applications, 36*(2), 2592–2602. doi:10.1016/j.eswa.2008.02.021

Office of the Communications Authority. (2015). Table 3: Telecommunications Services. In *Key Communications Statistics*. Retrieved from http://www.ofca.gov.hk/en/media_focus/data_statistics/key_stat/

Oliver, R. L. (1997). *Satisfaction: A behavioral perspective on the customer*. New York: Irwin McGraw Hill.

Oliver, R. L. (1999). Whence consumer loyalty? *Journal of Marketing, 63,* 33–45. doi:10.2307/1252099

Opsahl, T., Agneessens, F., & Skvoretz, J. (2010). Node centrality in weighted networks: Generalizing degree and shortest paths. *Social Networks, 32*(3), 245–251. doi:10.1016/j.socnet.2010.03.006

Osogami, T., & Harchol-Balter, M. (2006). Closed form solutions for mapping general distributions to quasi-minimal PH distributions. *Performance Evaluation, 63*(6), 524–552. doi:10.1016/j.peva.2005.06.002

Østerbø, O., & Grøndalen, O. (2012). Benefits of Self-Organizing Networks (SON) for mobile operators. *Journal of Computer Networks and Communications.*

Ouyang, Y., & Fallah, M. H. (2010). A performance analysis for UMTS packet switched network based on multivariate KPIs. In *Wireless Telecommunications Symposium (WTS),* 2010 (pp. 1-10). IEEE. doi:10.1109/WTS.2010.5479629

Ouyang, Y., Fallah, M. H., Hu, S., Yong, Y. R., Hu, Y., Lai, Z., & Lu, W. D. et al. (2014). A novel methodology of data analytics and modeling to evaluate LTE network performance. In *Wireless Telecommunications Symposium (WTS),* 2014 (pp. 1-10). IEEE.

Ouyang, Y., & Yan, T. (2015). Profiling wireless resource usage for mobile apps via crowdsourcing-based network analytics. *IEEE Internet of Things Journal, 2*(5), 391–398. doi:10.1109/JIOT.2015.2415522

Ouyang, Y., Yan, T., & Wang, G. (2015). CrowdMi: Scalable and diagnosable mobile voice quality assessment through wireless analytics. *IEEE Internet of Things Journal, 2*(4), 287–294. doi:10.1109/JIOT.2014.2387771

Package igraph, Version 0.7.1. (2014). Retrieved from http://igraph.org/2014/04/21/igraph-0.7.1-c.html

Panchenko, A., & Thümmler, A. (2007). Efficient phase-type fitting with aggregated traffic traces. *Performance Evaluation, 64*(7), 629–645. doi:10.1016/j.peva.2006.09.002

Patcha, A., & Park, J. M. (2007). An overview of anomaly detection techniques: Existing solutions and latest technological trends. *Computer Networks, 51*(12), 3448–3470. doi:10.1016/j.comnet.2007.02.001

**Compilation of References**

Peres, R., Muller, E., & Mahajan, V. (2010). Innovation diffusion and new product growth models: A critical review and research directions. *International Journal of Research in Marketing, 27*(2), 91–106. doi:10.1016/j.ijresmar.2009.12.012

Peyvandi, H., Imran, A., Imran, M. A., & Tafazolli, R. (2014, May). A target-following regime using Similarity Measure for Coverage and Capacity Optimization in Self-Organizing Cellular Networks with hot-spot. In *European Wireless 2014; 20th European Wireless Conference; Proceedings of* (pp. 1-6). VDE.

Peyvandi, H., Imran, M. A., & Tafazolli, R. (n.d.). *On Performance Optimization in Self-Organizing Networks using Enhanced Adaptive Simulated Annealing with Similarity Measure*. Academic Press.

Pham, H. (2007). *System software reliability*. Springer Science & Business Media.

Phua, C., Alahakoon, D., & Lee, V. (2004). Minority report in fraud detection: classification of skewed data. *ACM SIGKDD Explorations Newsletter, 6*(1), 50-59.

Pushpa, S. (2012). G.: An Efficient Method of Building the Telecom Social Network for Churn Prediction. *International Journal of Data Mining & Knowled Management Process, 2*(3), 31–39. doi:10.5121/ijdkp.2012.2304

Qian, H. (2011). China Mobile satisfaction survey analysis. *Jiangsu Science and Technology Information, 9*, 27–29.

Quinlan, J. R. (2014). *C4. 5: programs for machine learning*. Elsevier.

Rahmandad, H., & Sterman, J. (2008). Heterogeneity and network structure in the dynamics of diffusion: Comparing agent-based and differential equation models. *Management Science, 54*(5), 998–1014. doi:10.1287/mnsc.1070.0787

Ramaswamy, S., Rastogi, R., & Shim, K. (2000). Efficient algorithms for mining outliers from large data sets. In *ACM SIGMOD international conference on Management of data* (pp. 427–438). New York: ACM. doi:10.1145/342009.335437

Rawashdeh, A. (2015). Adoption of 4G mobile services from the female student's perspective: Case of Princess Nora University. *Malaysian Online Journal of Educational Technology, 3*(1), 12–27.

Reibman, A., & Trivedi, K. (1988). Numerical transient analysis of Markov models. *Computers & Operations Research, 15*(1), 19–36. doi:10.1016/0305-0548(88)90026-3

Reichheld, F. F. (2003). The one number you need to grow. *Harvard Business Review, 81*(12), 46–55. PMID:14712543

Reynolds, D. A. (1992). *A Gaussian Mixture Modeling Approach to Text-Independent Speaker Identification* (PhD thesis). Georgia Institute of Technology.

Ripley, B. D. (1996). *Pattern recognition and neural networks*. Cambridge, UK: Cambridge University Press. doi:10.1017/CBO9780511812651

Rogers, E. M. (1962). *The Diffusion of Innovations*. New York, NY: Free Press.

Rogers, E. M. (2010). *Diffusion of innovations*. Simon and Schuster.

Rossi, P. H., Wright, J. D., & Anderson, A. B. (Eds.). (2013). *Handbook of survey research*. Academic Press.

Sacerdote, B. (2001). Peer effects with random assignment: Results for dartmouth roommates. *The Quarterly Journal of Economics, 116*(2), 681–704. doi:10.1162/00335530151144131

Saeys, Y., Inza, I., & Larrañaga, P. (2007). A review of feature selection techniques in bioinformatics. *Bioinformatics (Oxford, England), 23*(19), 2507–2517. doi:10.1093/bioinformatics/btm344 PMID:17720704

Sakia, R. M. (1992). The Box-Cox transformation technique: A review. *The Statistician, 41*(2), 169–178. doi:10.2307/2348250

Saur, S., & Halbauer, H. (2011, May). Exploring the vertical dimension of dynamic beam steering. In *Multi-Carrier Systems & Solutions (MC-SS), 2011 8th International Workshop on* (pp. 1-5). IEEE. doi:10.1109/MC-SS.2011.5910725

Saylor, J. (2004, Oct 11). *The Missing Link in CRM: Customer Acquisition Management*. Retrieved from http://www.destinationcrm.com/Articles/Web-Exclusives/Viewpoints/The-Missing-Link-in-CRM-Customer-Acquisition-Management-44024.aspx

Schmittlein, D. C., & Mahajan, V. (1982). Maximum likelihood estimation for an innovation diffusion model of new product acceptance. *Marketing Science, 1*(1), 57–78. doi:10.1287/mksc.1.1.57

Seifi, N., Coldrey, M., & Svensson, T. (2012, December). Throughput optimization in MU-MIMO systems via exploiting BS antenna tilt. In *2012 IEEE Globecom Workshops* (pp. 653-657). IEEE. doi:10.1109/GLOCOMW.2012.6477651

Seifi, N., Coldrey, M., & Viberg, M. (2012). Throughput optimization for MISO interference channels via coordinated user-specific tilting. *IEEE Communications Letters*, *16*(8), 1248–1251. doi:10.1109/LCOMM.2012.060812.120756

Sequeira, K., & Zaki, M. (2002). ADMIT: Anomaly-based data mining for intrusions. In *8th ACM SIGKDD international conference on Knowledge discovery and data mining* (pp. 386-395). New York: ACM.

Sing, T., Sander, O., Beerenwinkel, N., & Lengauer, T. (2005). ROCR: Visualizing classifier performance in R. *Bioinformatics (Oxford, England)*, *21*(20), 3940–3941. doi:10.1093/bioinformatics/bti623 PMID:16096348

Snijders, T., Steglich, C., & Schweinberger, M. (2007). Modeling the coevolution of networks and behavior. Academic Press.

Snijders, T. A. (1996). Stochastic actor - oriented models for network change. *The Journal of Mathematical Sociology*, *21*(1-2), 149–172. doi:10.1080/0022225 0X.1996.9990178

Snijders, T. A. (2001). The statistical evaluation of social network dynamics. *Sociological Methodology*, *31*(1), 361–395. doi:10.1111/0081-1750.00099

Spark Programming Guide – Spark 1.6.0 Documentation. (2015). Retrieved from https://spark.apache.org/docs/1.6.0/programming-guide.html

Srinivasan, V., & Mason, C. H. (1986). Technical note-nonlinear least squares estimation of new product diffusion models. *Marketing Science*, *5*(2), 169–178. doi:10.1287/mksc.5.2.169

Subramanian, G. H. (1994). A Replication of Perceived Usefulness and Perceived Usefulness and Perceived Ease of Use Measurement. *Decision Sciences*, *25*(5-6), 863–874. doi:10.1111/j.1540-5915.1994.tb01873.x

Suykens, J. A. K., & Vandewalle, J. (1999). Least squares support vector machine classifiers. *Neural Processing Letters*, *9*(3), 293–300. doi:10.1023/A:1018628609742

Szlovencsak, A., Godor, I., Harmatos, J., & Cinkler, T. (2002). Planning reliable UMTS terrestrial access networks. *Communications Magazine, IEEE*, *40*(1), 66–72. doi:10.1109/35.978051

Tax, D. M. J., & Duin, R. P. W. (1998). Outlier detection using classifier instability. In *Advances in Pattern Recognition* (pp. 593–601). Berlin: Springer. doi:10.1007/BFb0033283

Theodoridis, S., & Koutroumbas, K. (n.d.). *Pattern recognition* (3rd ed.). Academic Press.

Tipper, D., Charnsripinyo, C., Shin, H., & Dahlberg, T. (2002, January). Survivability analysis for mobile cellular networks.*Communication Networks and Distributed Systems Modeling and Simulation Conference*, 367-377.

Trivedi, K. S. (2008). *Probability & statistics with reliability, queuing and computer science applications*. John Wiley & Sons.

Tsai, C. F., & Lu, Y. H. (2009). Customer churn prediction by hybrid neural networks. *Expert Systems with Applications*, *36*(10), 12547–12553. doi:10.1016/j.eswa.2009.05.032

Tsao, S., & Lin, C. (2002). Design and evaluation of UMTS-WLAN interworking strategies. In *Vehicular Technology Conference, 2002. Proceedings. VTC 2002-Fall. 2002 IEEE 56th* (Vol. 2, pp. 777-781). IEEE. doi:10.1109/VETECF.2002.1040705

Ultsch, A. (2002). Emergent self-organising feature maps used for prediction and prevention of churn in mobile phone markets. *Journal of Targeting, Measurement and Analysis for Marketing*, *10*(4), 314–324. doi:10.1057/palgrave.jt.5740056

Van den Bulte, C., & Lilien, G. L. (2001). Medical innovation revisited: Social contagion versus marketing effort. *American Journal of Sociology*, *106*(5), 1409–1435. doi:10.1086/320819

Van den Poel, D., & Lariviere, B. (2004). Customer attrition analysis for financial services using proportional hazard models. *European Journal of Operational Research*, *157*(1), 196–217. doi:10.1016/S0377-2217(03)00069-9

Viering, I., Dottling, M., & Lobinger, A. (2009, June). A mathematical perspective of self-optimizing wireless networks. In *2009 IEEE International Conference on Communications* (pp. 1-6). IEEE. doi:10.1109/ICC.2009.5198628

Von Neumann, J., & Burks, A. W. (1966). Theory of self-reproducing automata. *IEEE Transactions on Neural Networks*, *5*(1), 3–14.

Wang, W., Zhang, J., & Zhang, Q. (2013, April). Cooperative cell outage detection in self-organizing femtocell networks. In INFOCOM, 2013 Proceedings IEEE (pp. 782-790). IEEE. doi:10.1109/INFCOM.2013.6566865

Wang, W., Liao, Q., & Zhang, Q. (2014). COD: A cooperative cell outage detection architecture for self-organizing femtocell networks. *IEEE Transactions on Wireless Communications*, *13*(11), 6007–6014. doi:10.1109/TWC.2014.2360865

Wei, C. P., & Chiu, I. T. (2002). Turning telecommunications call details to churn prediction: A data mining approach. *Expert Systems with Applications*, *23*(2), 103–112. doi:10.1016/S0957-4174(02)00030-1

Wood, S. N. (2000). Modelling and smoothing parameter estimation with multiple quadratic penalties. *Journal of the Royal Statistical Society. Series B, Statistical Methodology*, *62*(2), 413–428. doi:10.1111/1467-9868.00240

Xie, L., Heegaard, P. E., & Jiang, Y. (2013, April). Network survivability under disaster propagation: Modeling and analysis. In 2013 IEEE Wireless Communications and Networking Conference (WCNC) (pp. 4730-4735). IEEE.

Yilmaz, O. N., Hamalainen, S., & Hamalainen, J. (2009, September). System level analysis of vertical sectorization for 3GPP LTE. In *2009 6th International Symposium on Wireless Communication Systems* (pp. 453-457). IEEE.

Yu, M. (2014). *China Mobile, customer satisfaction and promotion strategies for network companies* (Unpublished dissertation). Chongqing University, China.

Zhang, B., Cohen, W., Krackhardt, D., & Krishnan, R. (2011). Extracting Subpopulations From Large. *Social Networks*.

Zoha, A., Saeed, A., Imran, A., Imran, M. A., & Abu-Dayya, A. (2014, September). A SON solution for sleeping cell detection using low-dimensional embedding of MDT measurements. In *2014 IEEE 25th Annual International Symposium on Personal, Indoor, and Mobile Radio Communication (PIMRC)* (pp. 1626-1630). IEEE. doi:10.1109/PIMRC.2014.7136428

Zoha, A., Saeed, A., Imran, A., Imran, M. A., & Abu-Dayya, A. (2015, March). Data-driven analytics for automated cell outage detection in Self-Organizing Networks. In *Design of Reliable Communication Networks (DRCN), 2015 11th International Conference on the* (pp. 203-210). IEEE. doi:10.1109/DRCN.2015.7149014

# About the Contributors

**Chu (Ivy) Dang** is currently a doctoral candidate in Marketing at The Chinese University of Hong Kong (CUHK). She received a B.S. degree in Physics from Beijing Jiaotong University and a M.S. degree in Economics from CUHK. With training in both natural science and social science, she gained strong interest in quantitative modeling when entering the doctoral program in marketing. Her current research interests include social network effects, big data applications in marketing, and consumer behavior in e-commerce platform.

**Hasan Farooq** completed B.Sc. in Electrical Engineering from University of Engineering and Technology Lahore, Pakistan in 2009 and MS by Research on Networks from Universiti Teknologi Petronas, Malaysia in 2014. Currently, he is pursuing PhD in Electrical and Computer Engineering from University of Oklahoma, where his research focuses on Self-Organizing Networks (SON) for cellular systems. He is particularly focused on enabling self-healing in cellular system using machine learning and big data analytics techniques.

**Alexis Huet** is researcher and developer at the company Howso Technology, Nanjing, China. He received the Ph.D. degree in 2014 from University of Lyon, France. Most of his current research concerns computational statistics.

**Md Salik Parwez** completed B.Sc. in Electrical Engineering from University of Engineering and Technology Lahore, Pakistan in 2008 and ME in Information Science from Nara Institute of Science and Technology, Japan in 2014. Currently, he is pursuing PhD in Electrical and Computer Engineering from University of Oklahoma, where he is working on Self-Organizing Networks (SON) for cellular systems. Specifically, his research focuses on developing intelligent algorithms and frameworks using machine learning techniques and big data collected from cellular network. Additionally, he is directly involved in active projects on Wireless Virtualization for 5G Networks.

# Index

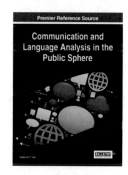

# Support Your Colleagues and Stay Current on the Latest Research Developments

# Become a Reviewer

In this competitive age of scholarly publishing, constructive and timely feedback significantly decreases the turn-around time of manuscripts from submission to acceptance, allowing the publication and discovery of progressive research at a much more expeditious rate.

**The overall success of a refereed journal is dependent on quality and timely reviews.**

Several IGI Global journals are currently seeking highly qualified experts in the field to fill vacancies on their respective editorial review boards. Reviewing manuscripts allows you to stay current on the latest developments in your field of research, while at the same time providing constructive feedback to your peers.

Reviewers are expected to write reviews in a timely, collegial, and constructive manner. All reviewers will begin their role on an ad-hoc basis for a period of one year, and upon successful completion of this term can be considered for full editorial review board status, with the potential for a subsequent promotion to Associate Editor.

Join this elite group by visiting the IGI Global journal webpage, and clicking on "**Become a Reviewer**".

**Applications may also be submitted online at:**
www.igi-global.com/journals/become-a-reviewer/.

Applicants must have a doctorate (or an equivalent degree) as well as publishing and reviewing experience.

If you have a colleague that may be interested in this opportunity, we encourage you to share this information with them.

**Any questions regarding this opportunity can be sent to:**
journaleditor@igi-global.com.

Printed in the United States
By Bookmasters